A Political Ecology of Women, Water, and Global Environmental Change

This edited volume explores how a feminist political ecology framework can bring fresh insights to the study of rural and urban livelihoods dependent on vulnerable rivers, lakes, watersheds, wetlands, and coastal environments. Bringing together political ecologists and feminist scholars from multiple disciplines, the book develops solution-oriented advances to theory, policy, and planning to tackle the complexity of these global environmental changes.

Using applied research on the contemporary management of groundwater, springs, rivers, lakes, watersheds, and coastal wetlands in Central and South Asia; Northern, Central, and Southern Africa; and South and North America, the authors draw on a variety of methodological perspectives and new theoretical approaches to demonstrate the importance of considering multiple layers of social difference as produced by and central to the effective governance and local management of water resources.

This unique collection employs a unifying feminist political ecology framework that emphasizes the ways that gender interacts with other social and geographical locations of water resource users. In doing so, the book further questions the normative gender discourses that underlie policies and practices surrounding rural and urban water management and climate change, water pollution, large-scale development and dams, water for crop and livestock production and processing, resource knowledge and expertise, and critical livelihood studies.

This book will be of interest to students and scholars of environmental studies, development studies, feminist and environmental geography, anthropology, sociology, environmental philosophy, public policy, planning, media studies, Latin American and other area studies, as well as women's and gender studies.

Stephanie Buechler is an Assistant Research Professor of Environmental Policy at the University of Arizona.

Anne-Marie Hanson is an Assistant Professor of Environmental Studies at the University of Illinois Springfield.

"Women around the world are facing the sharp edge of climate change impacts, especially impacts on water. So this truly feminist political ecology of gender, water and vulnerability is timely, important, and sobering. It's also exciting! From Lesotho to Mexico to Canada, justice and gender are on the agenda."

Paul Robbins, University of Wisconsin-Madison, USA

"This is an exciting example of how approaching environmental change through a gender lens usefully unsettles taken-for-granted boundaries. Placing people's experiences at the heart of the analysis, the book engages with recent insights from a wide range of disciplines to rethink global environmental change. It is indispensable reading for those interested in how water, social difference, and power interact to produce highly uneven waterscapes."

Margreet Zwarteveen, UNESCO-IHE Delft, The Netherlands

"This collection moves feminist political ecology forward in an unassuming and practical way. The authors open us to hidden connections between social inequalities and natural resource actions – especially how gender shapes our interactions with water in its many forms – interactions which in turn shape who we are as individuals and as societies."

Claudia Radel, Utah State University, USA

"Using multi-disciplinary analysis, this powerful collection illustrates how water intersects with women's everyday vulnerabilities. From North America to Central Asia, women's voices have shaped struggles over access to water, whether for agriculture or mining or domestic use in contexts of growing competition, conflict and increasing risks to climate change."

Sara Ahmed, International Development Research Centre, Asia Regional Office, India

"This is a thoughtful, theoretically, and methodologically timely book examining women's struggles and agency in light of rapid environmental change. The depth and complexity of the case studies provide compelling evidence of how gender shapes the manner in which water resources are contested; and how women's knowledge and actions lead to better understanding of the many factors shaping the relationship between people and natural resources in the Global South and North. This important edited volume will contribute significantly to the growing field of Feminist Political Ecology and will be of interest to a wide range of scholars in the social sciences and humanities."

Maria L. Cruz-Torres, Arizona State University, USA

Routledge International Studies of Women and Place
Series Editors: Janet Henshall Momsen and Janice Monk
University of California, Davis and University of Arizona, USA

1 **Gender, Migration and Domestic Service**
Edited by Janet Henshall Momsen

2 **Gender Politics in the Asia-Pacific Region**
Edited by Brenda S.A. Yeoh, Peggy Teo and Shirlena Huang

3 **Geographies of Women's Health**
Place, Diversity and Difference
Edited by Isabel Dyck, Nancy Davis Lewis and Sara McLafferty

4 **Gender, Migration and the Dual Career Household**
Irene Hardill

5 **Female Sex Trafficking in Asia**
The Resilience of Patriarchy in a Changing World
Vidyamali Samarasinghe

6 **Gender and Landscape**
Renegotiating the Moral Landscape
Edited by Lorraine Dowler, Josephine Carubia and Bonj Szczygiel

7 **Maternities**
Gender, Bodies and Spaces
Robyn Longhurst

8 **Gender and Family among Transnational Professionals**
Edited by Anne Coles and Anne-Meike Fechter

9 **Gender and Agrarian Reforms**
Susie Jacobs

10 **Gender and Rurality**
Lia Bryant and Barbara Pini

11 **Feminist Advocacy and Gender Equity in the Anglophone Caribbean**
Envisioning a Politics of Coalition
Michelle V. Rowley

12 **Women, Religion, and Space in China**
Islamic Mosques & Daoist Temples, Catholic Convents & Chinese Virgins
Maria Jaschok and Shui Jingjun

13 **Gender and Wildfire**
Landscapes of Uncertainty
Christine Eriksen

14 **Colonization and Domestic Service**
Historical and Contemporary Perspectives
Edited by Victoria K. Haskins and Claire Lowrie

15 **A Political Ecology of Women, Water and Global Environmental Change**
Edited by Stephanie Buechler and Anne-Marie Hanson

Also available in this series:

Full Circles: Geographies of Women over the Life Course
Edited by Cindi Katz and Janice Monk

'Viva': Women and Popular Protest in Latin America
Edited by Sarah A. Radcliffe and Sallie Westwood

Different Places, Different Voices: Gender and Development in Africa, Asia and Latin America
Edited by Janet Momsen and Vivian Kinnaird

Servicing the Middle Classes: Class, Gender and Waged Domestic Labour in Contemporary Britain
Nicky Gregson and Michelle Lowe

Women's Voices from the Rainforest
Janet Gabriel Townsend

Gender, Work and Space
Susan Hanson and Geraldine Pratt

Women and the Israeli Occupation
Edited by Tamar Mayer

Feminism / Postmodernism / Development
Edited by Marianne H. Marchand and Jane L. Parpart

Women of the European Union: The Politics of Work and Daily Life
Edited by Maria Dolors Garcia Ramon and Janice Monk

Who Will Mind the Baby? Geographies of Childcare and Working Mothers
Edited by Kim England

Feminist Political Ecology: Global Issues and Local Experience
Edited by Dianne Rocheleau, Barbara Thomas-Slayter, and Esther Wangari

Women Divided: Gender, Religion and Politics in Northern Ireland
Rosemary Sales

Women's Lifeworlds: Women's Narratives on Shaping their Realities
Edited by Edith Sizoo

Gender, Planning and Human Rights
Edited by Tovi Fenster

Gender, Ethnicity and Place: Women and Identity in Guyana
Linda Peake and D. Alissa Trotz

A Political Ecology of Women, Water, and Global Environmental Change

Edited by Stephanie Buechler
and Anne-Marie Hanson

Routledge
Taylor & Francis Group
LONDON AND NEW YORK

First published 2015
by Routledge
4 Park Square, Milton Park, Abingdon, Oxon OX14 4RN
605 Third Avenue, New York, NY 10017

Routledge is an imprint of the Taylor & Francis Group, an informa business

British Library Cataloguing-in-Publication Data
A catalogue record for this book is available from the British Library

Library of Congress Cataloging-in-Publication Data
A political ecology of women, water and global environmental change / edited by Stephanie Buechler and Anne-Marie S. Hanson.
pages cm.—(Routledge international studies of women and place)
ISBN 978-0-415-74935-0 (hardback) — ISBN 978-1-315-79620-8 (ebook) 1. Ecofeminism—Cross-cultural studies. 2. Water-supply—Social aspects—Cross-cultural studies. 3. Global environmental change—Social aspects—Cross-cultural studies. 4. Women in development—Cross-cultural studies. 5. Women in conservation of natural resources—Cross-cultural studies. 6. Political ecology—Cross-cultural studies. I. Buechler, Stephanie. II. Hanson, Anne-Marie S.
HQ1194.P65 2015
305.42—dc23 2014034250

ISBN: 978-0-415-74935-0 (hbk)
ISBN: 978-1-315-79620-8 (ebk)

Typeset in Goudy
by Keystroke, Station Road, Codsall, Wolverhampton

Contents

Figures

Tables

Contributors

Yvonne A. Braun, Associate Professor, Departments of International Studies and Women's and Gender Studies, and Director of African Studies, University of Oregon. Dr Braun teaches and advises students in the areas of social and environmental justice, development, and globalization. Her research includes over 15 years of ethnographic studies on Lesotho and Southern Africa. She has published widely in the areas of gender, water, development, inequality, intersectionality, and environmental justice.

Stephanie Buechler, Assistant Research Professor of Environmental Policy, Udall Center for Studies in Public Policy and School of Geography and Development, University of Arizona. She is an ambassador for Oxfam's 'Sister on the Planet' gender, development, and climate change initiative, and a former gender researcher for the International Water Management Institute. Dr Buechler's research focuses on the linkages between gender, water, climate change, and adaptation in rural and urban locations, urban, peri-urban, and rural agricultural production and processing, migration, livelihoods, and urban and rural development in semi-arid regions of Latin America, Asia, and North America.

Miriam Gay-Antaki, PhD student, School of Geography and Development, University of Arizona. Her research focuses on gender and climate change, feminist political ecology, and socio-environment relations in Mexico, her country of origin. She uses a feminist political ecology perspective to address carbon offset projects in Oaxaca and explores creative and more encompassing ways to engage women and other disadvantaged groups in climate change discourses, practices, and policies.

Ivan Golovnev, Documentary Filmmaker, specializing in anthropological subject matter. Ivan studied History at Omsk State University, and Film Direction at Sverdlovsk Film Studio in Yekaterinburg and "Higher Courses of Screenwriters and Directors" (VKSR, Russian – BKCP) in Moscow, under the instruction of film director Pyotr Todorovsky and screenwriter Natalya Ryazantseva. He has made many films focused on disappearing indigenous cultures within the Russian Federation. These films have been featured at many film festivals, receiving prizes on multiple occasions, in Russia and internationally.

Anne-Marie Hanson, Assistant Professor of Environmental Studies, University of Illinois Springfield. Dr Hanson teaches and advises students in political ecology, gender and natural resources, physical and human geography, and critical development studies. Her research has focused on oral histories of urban and coastal environmental change, eco-tourism, and sustainable development in and around natural protected areas of Mexico, Brazil, and the U.S. She works collaboratively with students, grassroots women's groups, and environmental agencies to address issues of garbage, health, and wetlands conservation in cities and coastal areas.

Leila M. Harris, Associate Professor, Institute for Resources, Environment and Sustainability and Institute for Gender, Race, Sexuality and Social Justice, University of British Columbia. Dr Harris is co-Director of the Program on Water Governance. Specific themes of her work include performativity of gender, ethnicity and difference/inequality, state-society relations, and the politics of scale in environmental change processes in southeastern Turkey, with First Nations in British Columbia, and in underserved settlements in Ghana and South Africa.

Eleanor Hayman, PhD candidate, Ludwig-Maximilian's University and affiliate at the Rachel Carson Center for Environment & Society, Munich, Germany. Eleanor has specific interests in the anthropology of water, environmental philosophy, traditional ecological knowledges, and ecological feminism. She is working collaboratively with citizens of Carcross/Tagish First Nation (CTFN), Yukon Territory, Canada on a framework with which to develop legislation for a CTFN Water Act rooted in Tlingit indigenous philosophy. This would set a precedent in Canadian law making.

Colleen James, *G̲ooch Tláa* (Dak̲l'aweidí Clan, Wolf Moiety, Carcross/Tagish First Nation). Colleen is a Carcross/Tagish First Nation General Council member and is an active participant in the Yukon River Inter-Tribal Watershed Council summits. She lives on the shores of Nares Lake near Carcross and is a steward of the land and waters. Colleen grew up with her grandmother Lucy Wren and is herself grandmother to 14 grandchildren.

Kathleen Kambic, Lecturer, Environmental Design Program, University of Colorado Boulder. Kambic holds a Master's degree in both Architecture and Landscape Architecture. She teaches courses on landscape architecture, architecture, and urban planning, including the seminar "Water Measure" on contamination, access and legibility of urban water infrastructure. Kathleen's research interests include urban water decline, site works, and hydrologic materials research, specifically addressing the relationship of water infrastructure and marginalized urban space and citizens.

Kuntala Lahiri-Dutt, Senior Fellow, Resource Environment and Development Program and Crawford School of Public Policy, College of Asia and the Pacific, Australian National University. Dr Lahiri-Dutt convenes the gender specialization in Master's of Applied Anthropology and Participatory Development and teaches courses on gender, fieldwork methods, and resource-environment-development. She is an editor of *ACME: An International E-Journal for Critical Geographies*. Her core area of research is gender and community in natural resource management in which she uses the lenses of social and environmental justice.

Diana Liverman, Regents Professor of Geography and Development and co-Director of the Institute of the Environment, University of Arizona and a visiting fellow with the Environmental Change Institute, University of Oxford. Dr Liverman is a renowned scholar on climate governance, the new carbon economy, and the impacts of climate on society. She has chaired and served on the USNRC Committee on the Human Dimensions of Global Environmental Change, the USNRC Committee on America's Climate Choices, the Science Advisory Committee for the Inter American Institute (IAI) for Global Change, and the International Council for Science (ICSU) Transition Team to create the new Future Earth program.

Andrea Ferreira Jacques de Moraes, Postdoctoral fellow in the School of Nutrition, Ryerson University, Canada. Dr Moraes also teaches for the Certificate in Food Security at the G. Raymond Chang School of Continuing Education at Ryerson. She is a member of the board of directors of CALACS (Canadian Association for Latin American and Caribbean Studies). Previously, she was coordinator for a Canadian International Development Agency funded development project on civil society participation in water management in Brazil. Her academic interests include gender and development, social justice, participatory democracy, and water and food security in Latin America.

Dina Najjar, Associate Social and Gender Scientist, the International Center for Agricultural Research in Dry Areas (ICARDA), a Consultative Group on International Agricultural Research (CGIAR) center. Her academic training included agricultural sciences, natural resource management, and socio-cultural anthropology. Dina's research interests include agricultural extension, rural women's empowerment and equality, women's access to and control over land, gender in agricultural innovation, and the consequences of the recent revolution in rural Egypt. She is currently working on gender and empowerment through agricultural innovations in Egypt, Tunisia, Morocco, Jordan, and Uzbekistan.

Patricia E. (Ellie) Perkins, Professor, Faculty of Environmental Studies, York University, Canada. Dr Perkins teaches and advises students in ecological economics, development, and critical interdisciplinary research design. Her research includes feminist ecological economics, climate justice, and participatory environmental education for political engagement. She directed research projects on community-based watershed education in Brazil and Canada, and on climate justice in watershed management with partners in

Mozambique, South Africa, and Kenya. She is a board member for the Green Change Project, focusing on green community development in Toronto.

Patricia Figueiredo Walker, Master in Environmental Studies (MES) candidate, Faculty of Environmental Studies, York University, Canada. Her research focuses on the vulnerability, adaptive capacity, and adaptation of poor and politically marginalized groups (especially women, children, and youth) to climate change. Previously Patricia coordinated a three-year International Development Research Center funded project on Climate Change Adaptation in Africa (CCAA).

Mark Wedge, *Aan Gooshú* (Deisheetaan Clan, Crow Moiety, Carcross/Tagish First Nation). Mark grew up in Carcross and currently lives in Tagish. He is a former *K̲áa Shaadé Héni* (Chief) of Carcross/Tagish First Nation (CTFN) and was the primary negotiator for CTFN's Final Agreement with the Canadian Government. He has acted as Executive Director of the Council of Yukon Indians and currently provides consulting and training services to governments and organizations in both Canada and the U.S.

Citt Williams, Documentary Filmmaker and Climate Scientist at the United Nations University, Tokyo. She has an MA in Documentary Arts and an MSc from the Climate Research Unit, University of East Anglia and the Oxford Internet Institute, University of Oxford. Her produced documentaries have been screened at prestigious festivals including Cannes and by broadcasters including Discovery and National Geographic, Canadian Broadcasting Corporation, and ABCTV. Her work encompasses intergovernmental, private sector, and community issues, and strives to build bridges with various partnerships through media content that entertains, educates, and inspires (http://ourworld.unu.edu).

Foreword

A quarter century of knowledge and change: pushing feminism, politics, and ecology in new directions with feminist political ecology

Leila M. Harris

Climate change, floods and drought, and water pollution are common headlines in the daily news. Just next to those stories, we might see reporting on gang rapes, unequal marriage rights based on sexual orientation, or even mass kidnapping of hundreds of girls participating in school examinations. These events appear to be disparate and unconnected, even if many readers might register these types of stories as similar in that they all characterize the negative 'crisis' headlines that so often dominate the media. Coming from a feminist political ecology (FPE) perspective, however, it is possible to understand that these issues may be connected—and indeed often are. The growing body of work on FPE has enabled us to understand that at the core of access to resources, knowledge of environments, or vulnerabilities to environmental changes are often key operations of social and political difference—including gender and sexuality, as well as ethnicity, race, poverty, or other inequalities. The same dynamics that drive unequal access to irrigation and drinking water or uneven participation in water user groups are those that also drive violence against women, differential treatment of 'minorities,' or changing norms and laws related to who can inherit property. It is only with this appreciation that we can understand why many indigenous women have participated in water walks around the Great Lakes of North America to bring attention to a range of freshwater concerns, or why thousands of women have protested the involvement of transnational companies in water provision in contexts such as India or Bolivia.

Nearly 25 years after the first studies we might recognize as FPE began to appear, work in this volume shows us the ways that the key aspects—those related to feminism, politics, and ecology—are being pushed in novel and insightful directions. Alongside other key contributions, the current volume is clearly pushing the boundaries of inquiry, and of understanding. For instance, we see moves to push what we mean by feminism towards more intersectional approaches, as well as analyses that also emphasize decoloniality, postcoloniality, and challenges to Eurocentric knowledges (e.g. Hayman, Williams and Golovnev, and Kambic, this volume). As such, these interventions also highlight key challenges to ways of knowing, for instance, emphasizing the importance of embodied and everyday lived experiences (e.g. Hanson, this volume), storytelling and narrative (Hayman and Hanson, this volume), or visual methods such as collaborative video (Williams and Golovnev, this volume). Doing so, the authors show, is important to be able to understand the gendered, intersectional, and justice effects of complex water-related changes, whether those associated with large dam building (Braun, this volume), or differentiated adaptation strategies in the face of climate change (Najjar, Buechler, and Perkins and Walker, this volume). The chapters here deploy and further these conceptual bases, challenging and advancing the fundamental concepts and analytical bases of FPE.

Pushing at the boundaries of 'feminism,' the chapters here build on the clear and compelling case related to the need to elaborate and understand women's knowledges or other gender-differentiated processes and effects, but also to consider other key intersectional dimensions of difference and inequality, whether ethnicity, race, poverty, or coloniality. The contribution by Hayman summarizes this imperative aptly in noting that FPE has the potential to unsettle and challenge dominant assumptions, particularly those that contribute to various ecological 'crises.' Yet, to do so, the very understanding of feminism itself must be problematized and unsettled, moving towards intersectional understandings (see the recent contribution by Mollett and Faria, 2013). For the case of dam building in Lesotho, this allows us to understand not only that women's livelihoods are affected by key changes associated with the projects, but also that it is the livelihood of certain populations (in relation to geography, ethnicity, and so on) that are sacrificed for the 'greater good' assumed by national developmentalist frameworks (Braun, this volume). For the context of participatory water governance in Brazil, an intersectional appreciation enables an appreciation of the fact that it is Afro-descended women who face particular difficulty in participating in governance institutions—despite the participatory rhetoric of Brazilian law (Moraes, this volume).

Pushing at the borders of what we might mean by 'politics' in FPE, it is increasingly apparent that novel approaches or alliances to get us beyond a number of current 'crises' will require new types of thinking, new interventions, and novel strategies. For instance, efforts to encourage climate change adaptation will be futile without a deep understanding of the ways that women's knowledges and livelihoods are affected in distinct ways—resulting in distinct adaptation strategies and livelihood practices (Buechler, this volume). As well, in order to evaluate the potential for water-related rights claims in a context such as Jharkhand state, in eastern India, Lahiri-Dutt (this volume) shows that we need to consider the ongoing and necessary tensions between informal and customary rights claims (for instance, those often made by women), and the more formal legal rights bases (including those that enable the ongoing extraction, use, or water pollution by the mining industry). Unpacking political and institutional bases and biases is necessary to confront ongoing environmental degradation, and also to foster more nuanced scholarship on key social and ecological challenges. Connecting to politics more broadly, and highlighting diverse activisms in evidence, it is also with the understandings offered by FPE that it becomes more possible to understand why women are so often at the forefront of water-related activism. Kambic (this volume) pushes these types of insights by suggesting that it is only through FPE and similar approaches that seek to contest control and make visible (hidden) uses of water in cities such as Los Angeles that we can attempt to promote alternative uses of urban space, or to begin to offer alternatives to the current uses and abuses of water and other key resources. Thinking about politics in terms of the everyday and embodied interactions, Hanson (this volume) illumines that women hold the potential and capacity to offer solutions to some global scale problems.

Finally, pushing 'ecology' in novel directions is increasingly an effort and intention of works in political ecology more generally. Buechler's contribution speaks to this imperative, highlighting the necessary conjoining of an FPE approach that highlights gendered adaptation strategies and climate change related effects, together with a socio-ecological systems approach that traces the contours and features of the changing biophysical environment. This is central, she suggests, to any effort to consider agricultural change, across the production cycle, and to explore the full range of possible adaptation strategies. Similar themes are picked up by Najjar to analyze women's and men's differentiated farming practices, with evidence that women utilize alternative water sources, and that women's crops in Egypt, although not supported, may be more environmentally sustainable than men's—both with clear implications for climate change adaptation, vulnerabilities, and agricultural policy. Also pushing notions of 'ecology' further, we can think about the multi-scalar dimensions of ecology and the ways that an explicit ecological focus invites attention to the ways that local environmental changes are necessarily linked to ongoing changes across scales. In the example from Mexico offered by Hanson, we see women's local activism and local conservation challenges in the Yucatán are necessarily linked to regional and global processes. These understandings are furthered by continuing to highlight not only the gender and feminist bases of FPE, but also by taking seriously the ecological and biophysical processes that are essential to delineate these connections.

For those who work in FPE, in many ways a volume such as this represents the maturation of the subfield. It is a welcome and useful addition to have an entire volume focused on feminist political ecologies of women, water, and global environmental change. While there had long been a sense that work along these lines was 'nascent,' a quarter century after the first moves in this direction the work now has a different feel—one that suggests that the conceptual bases are now well established, even if there is still considerable work to be done to continue extending and building FPE's analytical tools and frameworks. In addition to pushing at the key concepts that are the touch points of FPE, several other key insights are offered here. Specifically, in the pages that follow, we find work that argues for further rapprochement between feminist political ecology and environmental justice—two allied approaches that have remained somewhat distinct given their intellectual genealogy, but that clearly share similar goals and thematic interests. In addition, the chapters demonstrate that even with considerable progress to overcome gender-based inequalities, there remains key silencing of women's voices and needs with respect to water resource issues in many parts of the globe. For instance, work on mining in India (Lahiri-Dutt, this volume) highlights women's exclusion from water access, as well as from participatory processes that lead to further gender-based marginalization. These processes, in turn, foment considerable resistance among women. Focusing on women's involvement in water governance in Brazil, the contribution by Moraes also suggests that while progress may be observable for women's involvement in local water management, this does not always scale up to regional watershed committees. Our approach, therefore, to fostering gender inclusive participation must attend to the different norms and expectations at different levels of government, including the subtle differences that operate between them. Women's varied expertise on local ecological and social conditions, and on the impacts of climate change, represent an important fount of resilience for low-income communities. The examples cited by Perkins and Walker show how multi-scaled networking can foster the inclusion of women as valued experts in water and climate governance institutions, beginning at the local level.

Linked to knowledge production, method is yet another broad theme that emerges from across the chapters. Regarding knowledge production, there is particular interest in what might be gained through attention to diverse voices, or to foster alternative methods and modes of knowledge generation (e.g. attending to women's voices and embodied experiences, whether for waste issues, Hanson, or for climate change adaptation, Buechler and Williams and Golovnev). Of interest as well, works in this volume pick up on a growing interest in FPE, and across the social sciences and humanities, in arts based engagement and practice. Several chapters deal head on with these issues, confronting issues of power and inequality through video and visual methods, as well as through storytelling and involving alternative readings of environments, and, indeed, of the category of 'water' itself. Personally, I find some of this work in FPE to be among the most provocative and exciting. It presents opportunities to reimagine and extend the influence of our work to diverse publics, both in policy arenas and social change initiatives. The issue of access to knowledge and information is one that has long occupied feminists, and here, we see creative and effective responses to this challenge in ways that might extend and open up the debate on key socio-ecological challenges that we face. Tying the issues of alternative and pluriversal knowledges to ongoing environmental challenges, several contributions here clearly suggest that if we are to be successful at fostering alternatives to hegemonic (and tired or harmful) ways of doing things, we need to foster and think through alternative knowledges. To this end, the insights offered by Kambic on how to make visible hidden spaces, infrastructures, and uses of water are insightful, as are the discussions of narrative, storytelling, and indigenous knowledges (Hayman), collaborative video and visual methods (Williams and Golovnev), or new types of cross-regional collaborations to foster greater inclusion in water-related management (Perkins and Walker and Moraes). It is clear that a feminist and justice oriented project, or any meaningful engagement with sustainability, depends on these sorts of creative pathways to counter and recreate knowledges.

This volume weaves together exciting and innovative threads: building on feminist work in the past quarter century, we are able to forge avenues of engagement with diverse communities. It is increasingly clear that our feminisms—and ecologies—depend on such networks. No doubt these collaborations will continue to enrich our work in the quarter century to come.

Reference

Mollett, S. and C. Faria. 2013. Messing with gender in feminist political ecology. *Geoforum*, 45, 116–125.

Acknowledgments

This book was inspired by a double session organized by Stephanie Buechler and Anne-Marie Hanson at the Annual Association of Geographers (AAG) conference in April 2013 in Los Angeles, California and co-sponsored by the Coastal and Marine specialty group, Cultural and Political Ecology (CAPE), and the Geographical Perspectives on Women specialty group.

It is with gratitude that we acknowledge several people who supported this book project. First, we are grateful for the collaborative relationship born out of the work on the co-editorship of this book that transformed the relationship that was professor and graduate student into colleague and friend. We would also like to thank Jan Monk and Janet Momsen for their enthusiasm and insightful feedback on the chapters and on the overall project. We were blessed with wonderful contributing authors for this book project and feel very lucky to have had the opportunity to get to know each author through their contributions to this book. At any given time, emails flew around the world to Egypt, Brazil, several states in the U.S., Canada, Australia, England, Germany, India, and other locales! This project facilitated a discussion of women, water, and global environmental change that was truly international in scope, yet at the same time, profoundly local and based on critical engagement with women, men, and children in their communities.

It is with sincere thanks that we acknowledge the University of Arizona Udall Center for Studies in Public Policy's financial support for several parts of the production of the book and Robert Merideth of the Udall Center for providing editorial support on the reference list and citations. Editorial assistant Bethany Wright of Routledge provided critical and timely feedback at each stage of the production process.

We also wish to thank Alfredo Ribeiro of UFPE (the Federal University of Pernambuco, Brazil), Ludmilla Duarte Santana of Codevasf (government office of the São Francisco and Parnaíba valley water development authority in Brazil), Damon Lynch of the Anthropology Department at the University of Minnesota, François Molle of France and International Water Management Institute (IWMI), and Nitish Priyadarshi, independent Geologist, Ranchi, India for their beautiful photos for several chapters in this book. We are grateful for the help of Gary Christopherson at the School of Geography and Development, University of Arizona and Shipeng Sun at the University of Illinois, Springfield in the design of maps for two chapters in this book.

Stephanie is thankful first and foremost to the local women and men in the multiple regions where she has conducted ethnographic research over the years who have provided very valuable and rigorous training in women, water, and global environmental change. She would also like to thank her father, Hans Buechler, and her sister, Simone Buechler, for their guidance in conducting in-depth qualitative research through their incredible examples in Latin America, Europe, and the U.S. She is grateful for her mother's mentorship as an anthropologist focused on feminist research; she knows that if her mother were alive today that she would have enjoyed the contributions of the many authors in this volume. Stephanie's daughter Celina, a budding feminist and fervent environmentalist, provided insightful feedback on some of the chapters and contributed her poem written many years before that captured youthful awe of the animals, plants, and water in nature and human relationships with their environment. Husband Chris inspired love for all things water, and a passion that has led to deep and often long-lasting involvement with rural and urban communities around the world. Son Gabriel's volunteership at the Arizona-Sonora Desert Museum has offered dynamic learning opportunities through the lens of his young eyes on the most biodiverse of this planet's deserts and the crucial role that each drop of water plays there.

Anne-Marie would like to express her thanks to the women of coastal Yucatán who shared their time, their wisdom, and their friendship. She is deeply grateful for the many years of guidance and intellectual inspiration she received from mentors, colleagues, and students while at the University of Arizona, the Universidad Autónoma de Yucatán, and Trinity College. She sends a special thanks to Stephanie Buechler, Sarah Moore, Paul Robbins, Keiron Bailey, Diana Liverman, and Zayde Antrim, and to her other supportive colleagues who are too numerous to mention. In addition to intellectual guidance, her many favorite (wink) relatives and friends in Minnesota and elsewhere include a strong network of love and support wherever her work and life experiences take her. Her husband Manuel shared much needed humor and love while conducting the fieldwork that informs her chapter, and while adjusting to new places and career paths. Her siblings Rolf, Rachel, and Abe offered encouraging words, insights, and laughs when she most needed them. Parents Linda and Rolf Hanson provided her with moral support during a time of many transitions, including, at the end of the project, her own adventure with parenting. . . . And finally, to Lucas, she just wants to say: you're my world, kid.

1 Introduction

Towards a feminist political ecology of women, global change, and vulnerable waterscapes

Anne-Marie Hanson and Stephanie Buechler

Water flowing, never ceasing
Always on the move

Living and thriving
Frogs leap, fish dive,
The current's alive too!

Humans just pass it by,
Not giving a second glance,
But a river is fantastic,
You just have to give it a second chance

You can sit by the running water,
Magnificent sights to see,
Like a sapphire colored dragonfly,
Enjoying the morning breeze

There might be a javalina,
Having a sip to drink
Please don't scare him away
The river is his too!

We humans are too greedy,
All the water we use up,
Is enough for a pack of dogs,
And we don't seem to care!

We need to preserve water,
Or all the animals will leave,
Then we'll be all alone,
What a horrible place it would be!
Celina Scott-Buechler (2006; age ten)

On May 15, 2014, the Colorado River reached the ocean in the Gulf of California for the first time in over 20 years. A major drinking water source for over 40 million people in the western United States, throughout its history the Colorado River has also irrigated farmlands, supplied energy and resources for mining and urban centers, and enriched wildlife in a once-lush-but-now-desert Mexican delta. As part of an agreement and experiment between the United States and Mexico to restore riparian ecosystems in the delta, water was released from Lake Mead (Hoover Dam) to mimic the historic natural flooding that had halted due to a number of conditions (Postel 2014b). Decades of population growth, damming of the Colorado and other major rivers in the American Southwest, and climate change have contributed to water resource pressures across the southwest U.S. and northern Mexico (Postel 2014a; Waterman 2012). While the river reaching the sea seemed like a news-worthy triumph for the delta ecosystem, it will take serious changes in individual water consumption, community conservation efforts, as well as high-level policy and management changes to prevent future human-induced water scarcity in the southwest border region (Castle et al. 2014).

The poem above was written by a ten-year-old girl upon moving to Tucson, Arizona, who noticed the strange phenomena of rivers that either no longer flow at all or only flow when there is sufficient rainfall. A daughter of two environmental social scientists, she learned many things about water resources from living with them in semi-arid central Mexico and south India. Living in these water-scarce regions helped bring an appreciation of water that she carried with her to the southwest U.S. For example, she would partake (often by dancing outside) in the exuberance felt by all with the first rains of the North America monsoon in Tucson, the Asian monsoon in India, and the start to the rainy season in central Mexico. She took walks along streams or rivers with her family and with her classes at school in Tucson, thus becoming aware of dynamic ecological contexts that included human and animal life. Through her life experiences and a unit on water taught by her fifth grade teacher, she began to connect human activity with impacts on water bodies and on life that depended on those water bodies. This, in effect, was linking the political and the ecological, and it is a connection that can easily be grasped from a very young age. Time and again it is clear that this intimate connection is not fully understood or appreciated either by the general public, by policymakers or by scholars; yet the ramifications of this lack of understanding can be significant.

In particular in the U.S. southwest border region, 'ecological' concerns linked to global change are often under-publicized by the media when there are 'political' concerns related to immigration, border security, and drug violence in the same areas. Most recently, the many unaccompanied children who cross the desert areas of the U.S. southwest border to reunite with families and to seek asylum has focused attention on the political violence and poverty that are occurring in Central America. Only minor attention has been paid to gender issues in this migration. For example, in contrast to previous years when mainly men migrated from Central America, according to the Women's Refugee Commission, today nearly 50 percent of those migrants are girls, most between age five and 17 (de Silva Iddings 2014). Another recent media report documented a case of a Honduran mother and young daughter who took a loan from a drug trafficking gang to pay for their border crossing into the U.S. southwest, only to be caught and deported back to the danger and gang violence they were trying to escape (Stargardter and Palencia 2014).

Ecological issues that contribute to the rise in migration to the U.S. are rarely examined in this focus on child migration from Central America. The ecological issues are, however, linked with current economic and political conditions and include deforestation and intensified tropical storms. In 2007, Hurricane Felix affected 198,000 people in Honduras and Nicaragua, mainly in low-income Miskito Indian communities whose fishing grounds and agricultural fields were damaged (Nielsen 2010). In 1998, Hurricane Mitch caused almost a full year's precipitation to fall on most of Central America in a matter of days, and many rural and urban communities have never fully recovered. The lack of viable livelihood options has pushed some adults and children into business with criminal gangs and/or to migrate north in order to be able to have viable alternative livelihood options that are free from gang violence. Twenty thousand people were killed and two million displaced by Hurricane Mitch; most of those killed were in Honduras and, to a lesser extent, in Nicaragua and El Salvador (McLeman and Hunter 2010). As McLeman and Hunter explain:

> Decades of prior civil conflict in Nicaragua, Honduras, and El Salvador had created large populations of highly vulnerable people in the region, leaving many thousands of impoverished households occupying marginal and inherently hazardous lands in both the countryside and urban peripheries. These hazardous lands included steep slopes prone to failure and easily flooded low-lying areas. The vulnerability of these households was further exacerbated by ongoing deforestation and land degradation, and governments' inability to take measures to alleviate the endemic poverty.
>
> (2010:452)

This is contrasted with the case of Belize where Hurricane Mitch also struck, but with no lives lost, likely due to better environmental conservation programs in coastal areas, lack of previous civil conflict, and more progressive policies aimed at reducing social inequalities (ibid.).

As these examples indicate, ecological and political issues combine to significantly affect local populations, especially the poor. Those who are most vulnerable to poverty in Central America, including women and children (on urban Honduran boys see Schmidt and Buechler 2014), now comprise over half of the migrants to 'El Norte' in search of an economically and physically (bodily) secure future.

Gender, water, and global environmental change

The goals of our edited volume are to further explore these uneven and scalar relationships between political, social, and ecological systems through current international empirical research and to generate multidisciplinary scholarship in feminist political ecology on gender, water, and global environmental change. We focus on the practical implications of women's work and knowledge networks, and how these are linked to other forms of social differentiation as played out through the many forms of human relationships with water and global change. To the extent that people depend on water resources that are increasingly limited and linked to multinational political decisions and global climate shifts, this volume underscores the need for further attention to be placed on the particular political, environmental, economic, and social systems within which livable habitats are made and unmade.

People who are socially, economically, culturally, politically, institutionally, or otherwise marginalized are often highly vulnerable to the substantial and increasing impacts of environmental change (Romero-Lankao et al. 2014). Women and other marginalized groups in economically poor world regions are the most vulnerable to negative effects of environmental change, and yet, grassroots activist groups in these areas are often the most resilient and vocal proponents of protecting the environment at both local and global scales. We build on previous feminist political ecology research to recognize that, most often, so called 'minorities' by gender, race, class, and ethnicity are unfairly disadvantaged in the face of restricting political economies and climate extremes (Salleh 2009b).

The studies included in this volume illustrate some of the recent changes in political and ecological climates that have had severe consequences for natural and human systems on all continents and across the oceans (Romero-Lankao et al. 2014). Deterioration of water quality, rising temperatures, and changes in the seasonality, quantity, and duration of precipitation are combined issues that increasingly alter human, animal, and plant demand for water resources. At the same time, more extreme weather and climate events—including droughts, floods, hurricanes, and intense rainfall—further the impacts on economic sectors, resource-dependent livelihoods, food security, and human health (Alston 2013). In particular, extreme changes to surface water and coastal resources intersect with urbanization, agricultural intensification, and large-scale water use processes to intensify already existing social, economic, and environmental inequalities.

This volume calls attention to these inequalities as they are tightly linked to issues of women, water, and rapid environmental change. As Nelson and Stathers argue, "everyone plays a role in socio-ecological systems" (2009:65). For that reason, ecosystems and social systems must be examined in relation to each other, as both have different capacities and go through stages of adaptation to change (Holling et al. 2002). Resilience scholars influence our study of how natural resources will be impacted by global environmental change and "how localized environmental degradation, resource rights, and a changing climate will interact" (ibid.:63).

We focus on women not as unitary subjects, but as strategically important for future resilience and mitigation of global change issues as they are experienced in localized sites and through embodied subjects. Women generally experience worse environmental conditions than men for many reasons. Across all cultures, races, and ethnic groups, women are more likely to be poor than men, exposing them to environmental change and environmental problems endemic to poverty in rural and urban areas (Buckingham et al. 2005). Women frequently are the caretakers, health care providers, food provisioners, and small-scale farmers, and are responsible for domestic water provision and management (Bennett et al. 2005; Buechler 2009; Radel 2009). These roles are often 'naturalized,' unpaid, and under recognized, but also bring women into daily and direct contact with environmental pollution and resource scarcity and give access to issues linked to political and climate shifts (Salleh 2009b). Often children accompany women in their daily work, and women thus pass on knowledge about political and ecological conditions related to water resources to future generations. Organizations also can play important roles in water education, particularly in urban and rural communities that are unfairly disadvantaged by pollution, climate, and political change.

Throughout this book, we stress how women, from (and despite) their marginalized positions, are very active in their communities' efforts to mitigate the negative effects of environmental change on water resources. Women are also increasingly proactive in collaborative work with larger environmental activist networks and groups beyond the confines of their own communities to address human engineered water problems. Through several case studies, we move from a focus on women's vulnerabilities and the silencing of gendered and subaltern knowledges to stress the global significance of women's embodied practices, emerging social movements, and collaborative action (Escobar 2004; Gibson-Graham 2006; Gibson-Graham and Roelvink 2009).

As researchers from several different nations concerned with these issues, this volume is an attempt to bring together political ecologists and feminist scholars from multiple disciplines to share applied research on the contemporary management of rivers, lakes, watersheds, and coastal areas in Central and South Asia; Northern, Central, and Southern Africa; and South and North America. As scholars, practitioners, and activists, we use insights from feminist political ecology to focus on water bodies as sites where global environmental processes both produce and are greatly influenced by multiple forms of social difference. These processes occur within the particularities of local settings and include diverse understandings of gender, class, race, and ethnicity. As feminist scholars, we acknowledge that our perspectives are filtered through our own positionality as scholars with international and varied work experience but currently situated in universities or research institutes. We bring our own political sensitivities and critical theoretical lenses to explore questions of changing environmental conditions as well as access to and control over water resources for agriculture and agricultural processing, small-scale enterprises, and household use. We also examine alternative and gendered knowledge systems, innovations in urban design, and focus on multi-scalar grassroots activism led by women to address environmental change.

In the following sections of this introduction, we first address the broader role of political ecology in framing issues of poverty, resilience, and social justice as related to global environmental change. The second part explains our explicit focus on women and gendered subjects. We address how feminist political ecology privileges the knowledge of those most affected or marginalized by neoliberal, colonial, and patriarchal systems in which water and climate policy and practice are carried out. We consider the explicit connections between feminist methodologies and new directions in feminist political ecology to consider the uneven power relations through which women, water, and global environmental change processes occur and play out in the subsequent chapters that form the core of this collection. Finally, we describe the three main themes of the book and the main contributions of each chapter to policymaking and theory building.

Political ecology, social justice, and environmental change

Political processes at various scales influence water resource and livelihood issues. Climate mitigation responses are also closely connected to greater fluctuations in water quality and social, economic, and technical accessibility. In the last few decades, political ecology has emerged as one of the most overarching and critical interdisciplinary fields linking theorization of and practices pertaining to human-environment relations. In short, political ecology (PE) includes analysis of complex networks of relationships between humans and environments (Robbins 2004). Political ecological analysis often includes a focus on the linkages between social and environmental disparities that are experienced unequally among marginalized groups. By focusing on political issues of "access and control over resources and their implications for environmental health and sustainable livelihoods," political ecology has been adept at explaining environmental conflict in terms of struggles over "knowledge, power and practice" and "politics, justice and governance" (Watts 2000:257).

Throughout the field's history, however, there have been several criticisms in which scholars (from both within and/or outside the field) claim that PE is deficient in 'ecology' (Vayda and Walters 1999; Walker 2005) or lacking 'politics' (Bryant and Jarosz 2004; Keil 2003; Rocheleau et al. 1996). Blaikie has argued that PE scholars tend to overgeneralize and to under-play human agency and diversity (Blaikie 2012). In the area of PE and development policy, its discursive treatments tend to "over-emphasize representation at the expense of practice" (Blaikie 2012:234; Peet and Hartwick 2009). While strongly focused on representation and power relations that unevenly and disproportionately affect local environments and livelihoods within larger political and economic systems (Escobar 2008; Zimmerer and Basset 2003), until recently (see, for example, Ogden et al. 2013), PE also underplayed the concept of resilience, and the proactive actions of women and other marginalized groups that work to address adaptation challenges collectively and from the margins.

Socio-ecological resilience is the capacity of a system (such as an ecosystem or a livelihood system) to recover after a shock such as a flood or drought or a political or economic crisis (Adger 2000; Pickett et al. 2013). Resilience to global environmental change must also be understood in terms of the interplay of institutional, political, and economic processes that occur in specific, rooted locations (Rocheleau and Roth 2007), within complex assemblages and networks (Gibson-Graham and Roelvink 2009), and between scales. As Walker and Salt argue, "the scale in which we are interested in is connected to and affected by what's happening at the scales above and below, both in time and in space" (2006:88).

These critiques have led to a more recent focus in PE on the importance of interdisciplinarity as a central tenet of critical research design in human-environment relations (Blaikie 2012; Castree et al. 2010; Ogden et al. 2013). As Ogden et al. state, "the Anthropocene's heterogeneous socio-ecological processes sound an important warning about the need to seek alternative forms of thinking about, and action toward, the world around us" (2013:346). In particular, in the area of human-water research, multidisciplinary and 'engaged' research can explicitly link the social and ecological drivers of global change, and further explore the implications of these linkages for policy, science, environmental justice, and social action (Bakker 2010; Sultana and Loftus 2012). This resonates well with current goals of many feminist environmental scholars and political ecologists to create environmental justice at all scales (Agarwal 2010; Bickerstaff et al. 2009; Blaikie 2012; Buckingham and Kulcur 2009; Resurrección 2013; Sundberg 2008). PE, then, with the help of feminist analysis and a renewed focus on resilience, is moving in new directions towards interdisciplinarity; this volume is an example of this move.

Vulnerable waterscapes and feminist political ecology

A strong research focus on women and water is quite recent (Bennett et al. 2005; Buechler and Zapata 2000; Coles and Wallace 2005; Cruz-Torres 2008; Lahiri-Dutt 2006; Lynch 1991; O'Reilly 2006; Zwarteveen et al. 2012), but overall, the connections between women, water, and global environmental change are still understudied and frequently undervalued in academic, policy, and development literatures. Despite accumulated evidence that gender is a significant dimension in environmental sustainability and development, gendered analyses of socio-environmental issues continue to be viewed as 'special interest,' and their incorporation into program interventions has been piecemeal at best (Agarwal 2010). There is also a tendency to frame discussions of women around their vulnerabilities in relation to broader structural and environmental forces, rather than focus on women's agency and knowledge (Buechler n.d.; Hanson 2015; Resurrección 2013). Furthermore, "within environmental justice studies, although many campaigns signal the disproportionate impacts of environmental problems on women, gender inequalities are not yet prominent in environmental justice campaigns or their academic analyses" (Buckingham et al. 2005:428).

Feminist political ecology has allowed us to address how the same dynamics that produce unequal access to resources or disproportionate vulnerabilities to environmental changes are often key components of social and political difference. As a conceptual framework for examining human-environment issues, feminist political ecology was introduced in the 1996 Routledge volume edited by Rocheleau, Thomas-Slayter, and Wangari. That publication demonstrated the necessity for political ecologists to focus on gender as a critical variable in shaping socio-ecological change, resource access and control, and possibilities for sustainable development. As scholars and activists have shown, there is:

> limited evidence of gender differences in environmental attitudes and behaviors overall, but there is evidence that women and men differ in the types of environmental issues with which they engage politically, as well as in the manner in which they carry out their political and activist interests.
> (Radel 2009:334)

In this sense, feminist political ecology, including PE analysis and feminist methodologies, highlights the intersections between social identity (including gender identity), political and human-environment relations, and the specific material outcomes of those intersections (Radel 2012).

Political ecologists and feminist researchers tend to work in a "boundary zone between positivist and critical paradigms, consciously combing critical theory, empiricist fieldwork, and quantitative and qualitative analysis" (Rocheleau 1995:458). In terms of theory, feminist research has long addressed scales of analysis to include the interactions between intimate, household, and global-level practices and politics (Elmhirst 2011). As Mountz and Hyndman argue, "the global and the intimate constitute one another" (2006:446). Feminist scholars stress, for example, that frequently women act to connect women and men with the global economy, and to local and global institutions (Buechler 2012; Chakrabarti 2014; Nash 2005).

Feminist analysis is not solely focused on women, but on social and material inequalities more broadly. Feminist analysis often includes issues related to social and cultural practices in private and public spaces and how these spheres are integrally connected. Haraway argued that women have always been networked (2013), and as Rocheleau and Roth have commented, "We [political ecologists] have long been creators of networks that weave together markets, materials, money and ideas, with people, places, plants and animals in myriad constellations of constantly emerging social-ecological formations" (2007:433).

Indeed, PE does converge with feminist research on many points, particularly in addressing issues of scale, rights and responsibilities, and marginalization and vulnerability, but credit is rarely given to feminist theory or methodology as major contributors to this type of analysis (Elmhirst 2011). PE more broadly examines relationships of power, but studies frequently privilege class as the most important source of social-political inequality. Feminist political ecology explicitly addresses class issues of power, but also stresses that the same importance be given to other forms of power dynamics and social stratification such as gender, race, and ethnicity.

As such, feminist political ecology is gaining a renewed momentum by entering into dialogues with conceptual approaches that both broaden and deepen understanding of the intricate links between social difference and ecological change (Elmhirst 2011; Harris 2009; Rocheleau and Roth 2007). New theoretical approaches have added to previous feminist studies. Some of the most prominent examples of recent work include actor-network theory and intersectionality with relation to de-centered subjects and socio-environmental identities that are co-produced through power relationships linked to gender, race, caste, and environments (Bezner Kerr 2014; Mollett and Faria 2013; Nightingale 2011). Environmental justice frameworks have also strengthened the links of critical feminist political ecological analyses to their direct material and policy implications (Kimura and Katano 2014). Gendered positionality and empowerment continue to be themes that importantly connect not only private and public spaces, but also urban and rural socio-environmental networks (Hovorka 2006). Embodied feminisms and emotional geographies are used as frames to understand how human-water inequalities and subjectivities are created through and intersected by ecological change, spatial relations, and environmental imaginaries—all of which are linked to emotional and power-laden relationships (Harris 2014; Sultana 2009; Truelove 2011). Finally, ethical political ecologies draw from ethics of care—both of environments and towards social groups—to demonstrate how gendered human-environment relations often form part of post-capitalist politics and challenge neoliberal notions of dependent environmental subjects (Jarosz 2004, 2011; Salleh 2009a).

The new theorizations within recent feminist political ecology literature have created significant and vibrant opportunities for seeking multidisciplinary and multi-scalar perspectives on gendered environmental inequalities. While early feminist political ecology was explicit in its focus on mixed-methodologies as a way of expressing the multiple voices (Moss 2002; Nightingale 2003; Rocheleau 1995), like PE more broadly, there has not been as much explicit focus on methodologies in recent feminist political ecology literature. And with a few notable exceptions (for example, the European Network of Political Ecology, ENTITLE, or the University of Arizona's Public Political Ecology Lab), focus has not explicitly addressed how critical theoretical analysis within PE is and can be more directly applied towards interdisciplinary efforts to change policy outcomes. This volume seeks to address these gaps through a discussion in each chapter of the often innovative methodologies employed to conduct collaborative research with women and other socially marginalized groups who seek environmental justice, material outcomes, and policy changes.

We also center this volume on women and water, primarily to draw attention to the interplay between different types of water bodies and their relation to gendered roles and responses to environmental change. We find that feminist political ecology is a particularly productive framework to address these linkages that have been understudied in the environment and gender literature. We are advancing knowledge on gender and water and gender and climate change by adding to the small but growing literature dedicated to these topics and by linking them. Rather than shying away from a focus on women, we embrace this focus without advocating the ethnocentric, elitist, or essentialist eco-feminist views that women (as an unproblematized or generalized social category) are inherently closer to nature (Leach 2007). Instead, we focus on how men and women have different spaces in which they operate within intricate networks of social differentiation that vary across cultures (Gaard 2011).

The spaces where women operate are frequently less visible than the spaces of men's productive work. The water-related work that women do often remains invisible or under-recognized and women often do not have the same formalized privileges for access to water resources. Commonly, land rights are prerequisites to water rights and women are much less likely to inherit land or have the necessary capital to purchase land. They are often barred from decision-making concerning water resources for domestic use and for irrigation (Zwarteveen 1997; Zwarteveen et al. 2012).

As a result and despite gender mainstreaming efforts, women are repeatedly outside the central focus of water and climate policy and programs. This collection questions the normative gender discourses that underlie policies and practices surrounding water management and climate change, large-scale development and dams, resource knowledge and expertise, and critical livelihood studies. Chapters show linkages between coastal and inland areas; rising sea levels and desertification; water resources in and around lakes and rivers, springs and wells; diversions of water into dams, mines, and urban water supply networks, and the use of irrigation drainage water and urban greywater. Each of the authors in this volume also addresses the relationships intertwining social and ecological systems and the ways in which these relationships affect women's individual and collective strategies for dealing with water resource issues and environmental change. In addressing these linkages, chapters reveal women's attempts to achieve greater resiliency from the micro to the global scales.

Major themes and chapter contributions

Part I: Feminist political ecology and large-scale water resource management

In the first part of the book, chapters address large-scale water resource management projects and their local impacts in terms of social difference, environmental justice, and decision-making processes. Chapters examine the multiple layers of social difference (i.e. gender, race, ethnicity, social class, geographical location) that affect power relationships in large-scale water management. These uneven power relationships hinder efforts to address challenges in the face of climate change and resource extraction in urban and rural areas. We pursue a feminist political ecology critique of these large-scale water projects based on dominant environmental paradigms that overlook or obscure these power differences and that contribute heavily to uneven resource access and environmental decline.

Yvonne Braun combines intersectionality with environmental justice frameworks to understand environmental subjectivities, material-ecological inequalities and their relationship to large dams in rural and urban areas of Lesotho. Kuntala Lahiri-Dutt examines large-scale water management by coal mining companies in the state of Jharkhand in eastern India. These companies treat land and water rights together, and therefore she tells the story of indigenous women who fight for their rights to water by resisting the dumping of mining wastewater into community water sources and the diversion of their spring water to mining camps. Drawing from landscape architecture and feminist political ecology, Kathleen Kambic draws attention to the exclusionary spaces of urban water infrastructure and the orphaned (undervalued and unkempt) spaces of urban rivers in Los Angeles. She discusses how water infrastructure has become hidden as it is controlled through large-scale management from above. She explores women's small organizations' work to bring 'hidden' infrastructure back to public space. Andrea Ferreira Jacques de Moraes's chapter focuses on scalar issues of watershed management in Brazil's semi-arid region. She addresses how intersectionality issues related to gender, race, and class have severe implications for the active participation of citizens in managing water during times of scarcity and increasing women's representation in water resources issues in Northeastern Brazil.

Part II: Women and innovative adaptation to global environmental change

In climate change research, there is a tendency to frame discussions of women around their vulnerabilities rather than focus on women's agency and the ways in which women's knowledge is critical to understanding environmental degradation and climate change and central to creating innovative adaptation strategies. In this part, authors address the challenges and innovative adaptation strategies of women in the face of increased water scarcity and climate-related livelihood problems.

Stephanie Buechler explores the adaptive mechanisms of women cheese producers and male ranchers and dairy producers to climate change. They develop despite the fact that they are working in the excluded spaces of current water management policy in an ecologically fragile riparian area of northern Sonora, Mexico. Negative and positive tipping points are visible in the community. Patricia E. (Ellie) Perkins and Patricia Figueiredo Walker examine cross-regional and cross-watershed collaborations among Brazilian and Canadian partners and equity-focused climate change adaptation in a project with partners in Canada, Kenya, Mozambique, and South Africa. Dina Najjar focuses on two desert settlement areas in Egypt, which are part of the massive Mubarak Resettlement Scheme to expand the area farmed and to compensate for lands lost due to climate change-induced rising seas levels. Women are innovating through water use conservation and planting crops with low water requirements, but are not supported by government policies and programs.

Part III: Stories, narratives, and knowledge production of socio-environmental change

Stories help us make sense of the world. Rather than position women's words as subaltern testimonials that normalize colonial, patriarchal, and neoliberal state projects (Curiel 2007; Hanson 2015), chapters in this part focus on indigenous and women's stories as sources of critical knowledge for understanding and adapting to global environmental change. Stories in the three chapters were collected and shared using participatory feminist methodologies: ecological oral histories, indigenous narrative storytelling, collective Skype meetings and interviews, and collaborative filmmaking.

Anne-Marie Hanson draws from women's oral histories of socio-environmental change to address the gendered exclusionary spaces of nature protection and urbanization that are exacerbated by overflows of local and globally produced plastic wastes, rising sea levels, and more intense storms in the coastal wetlands of northwestern Yucatán, Mexico. She also illustrates women's activism in confronting waste-related challenges in coastal cities and tourist areas. Eleanor Hayman and her collaborators combine indigenous narrative storytelling with feminist and post-colonial analyses of climate change relationships through discussions of the water-based cosmologies of the Tagish and Tlingit peoples in the Yukon lakes region in Canada. They address the biases of western (colonial) land-based management and how these exclude water-centered cosmologies and their implications for future natural resource management. Through innovative media studies and feminist environmental analysis, Citt Williams and Ivan Golovnev work collaboratively with indigenous Pamiri women and national and international climate change scientists to address multiple knowledges of climate change and its importance for international negotiations, local women's agency, and more holistic understandings of climate change media communications.

This book is an extension of women's storytelling and political action from the intimate to the global scale. Women and indigenous peoples' voices are brought to the fore to highlight their critical roles in confronting environmental change that is affecting the water they, their households, their communities, and the larger ecosystems they are part of, depend on for material and physical well-being. The authors' innovative methodologies and the linking of theoretical frameworks help illuminate the action these groups have taken to address gendered water resource issues based on deep and dynamic knowledge. To offer insights into alternative ways to address complex challenges such as those discussed in the beginning of this chapter related to the U.S. southwest, the studies presented here pay careful attention to the interplay between ecological processes, political power dynamics among individuals, communities and regions, and "water flowing, never ceasing, always on the move."

References

Adger, W.N. 2000. Social and ecological resilience: are they related? *Progress in Human Geography*. 24 (3):347–364.

Agarwal, B. 2010. *Gender and green governance: The political economy of women's presence within and beyond community forestry*. Oxford and New York: Oxford University Press.

Alston, M. 2013. Introducing gender and climate change: Research, policy and action. In *Research, action and policy: Addressing the gendered impacts of climate change*, edited by M. Alston and K. Whittenbury, 3–14. Dordrecht, London, New York and Heidelberg: Springer.

Bakker, K. 2010. *Privatizing water: Governance failure and the world's urban water crisis*. Ithaca, NY: Cornell University Press.

Bennett, V., S. Dávila-Poblete, and N. Rico (eds). 2005. *Opposing currents: The politics of gender and water in Latin America*. Pittsburgh, PA: University of Pittsburgh Press.

Bezner Kerr, R. 2014. Lost and found crops: agrobiodiversity, indigenous knowledge, and a feminist political ecology of sorghum and finger millet in northern Malawi. *Annals of the Association of American Geographers*. 104 (3):577–593.

Bickerstaff, K., H. Bulkeley, and J. Painter. 2009. Justice, nature and the city. *International Journal of Urban and Regional Research*. 33 (3):591–600.

Blaikie, P. 2012. Should some political ecology be useful? The Inaugural Lecture for the Cultural and Political Ecology Specialty Group, Annual Meeting of the Association of American Geographers, April 2010. *Geoforum*. 43 (2):231–239.

Bryant, R.L. and L. Jarosz. 2004. Ethics in political ecology: a special issue of Political Geography. *Political Geography*. 23 (7):807–812. doi: 10.1016/j.polgeo.2004.05.010.

Buckingham, S. and R. Kulcur. 2009. Gendered geographies of environmental injustice. *Antipode*. 41 (4):659–683.

Buckingham, S., D. Reeves, and A. Batchelor. 2005. Wasting women: the environmental justice of including women in municipal waste management. *Local Environment: The International Journal of Justice and Sustainability*. 10 (4):427–444.

Buechler, S. 2009. Gender, water, and climate change in Sonora, Mexico: implications for policies and programmes on agricultural income-generation. *Gender & Development*. 17 (1):51–66. doi: 10.1080/13552070802696912.

——. 2012. Sustainability of gendered agricultural production in the context of water scarcity and climate change on the Mexican border with the United States. In *Gender and sustainability: Critical case studies from Asia and Latin America*, edited by M.L. Cruz Torres and P. McElwee, 121–141. Tucson, AZ: University of Arizona Press.

——. n.d. Grassroots initiatives to address climate-water vulnerabilities in Northwest Mexico.

Buechler, S. and E. Zapata. 2000. *Género y manejo de agua y tierra en comunidades rurales de México, serie Latinoamericana*. Mexico, D.F.: International Water Management Institute and the Colegio de Postgraduados.

Castle, S.L., B.F. Thomas, J.T. Reager, M. Rodell, S.C. Swenson, and J.S. Famiglietti. 2014. Groundwater depletion during drought threatens future water security of the Colorado River Basin. *Geophysical Research Letters*. Advance online publication. doi: 10.1002/2014GL061055.

Castree, N., P. Chatterton, N. Heynen, W. Larner, and M.W. Wright. 2010. Introduction: the point is to change it. *Antipode*. 41:1–9. doi: 10.1111/j.1467-8330.2009.00713.x.

Chakrabarti, S. 2014. *The gender advantage: Women on the front line of climate change*. Rome, Italy: International Fund for Agricultural Development. 21pp. www.ifad.org/climate/resources/advantage/gender.pdf. Accessed June 4, 2014.

Coles, A. and T. Wallace (eds). 2005. *Gender, water and development*. Vol. 27, *Cross-cultural perspectives on women*. Oxford: Berg Publishers.

Cruz-Torres, M.L. 2008. *Lives of dust and water: An anthropology of change and resistance in Northwestern Mexico*. Tucson, AZ: University of Arizona Press.

Curiel, O. 2007. Crítica poscolonial desde las prácticas políticas del feminismo antirracista. *Nómadas*. 26:92–101.

de Silva Iddings, A.C. 2014. "Girl immigrants crossing the US border often survived rape and abuse. They've got more courage than most of Congress." *The Guardian*, August 1.

Elmhirst, R. 2011. Introducing new feminist political ecologies. *Geoforum*. 42 (2):129–132. doi: 10.1016/j.geoforum.2011.01.006.

Escobar, A. 2004. Beyond the Third World: imperial globality, global coloniality and anti-globalisation social movements. *Third World Quarterly*. 25 (1):207–230.

——. 2008. *Territories of difference: Place, movements, life, redes*. Durham, NC: Duke University Press.

Gaard, G. 2011. Ecofeminism, revisited. *Feminist Formations*. 23 (2):26–53.

Gibson-Graham, J.K. 2006. *A postcapitalist politics*. Minneapolis, MN: University of Minnesota Press.

Gibson-Graham, J.K. and G. Roelvink. 2009. An economic ethics for the anthropocene. *Antipode*. 41 (1):320–346.

Hanson, A.-M. 2015. Women's oral histories of recycling and development in Celestún, Mexico. *Gender, Place & Culture*.

Haraway, D. 2013. *Simians, cyborgs and women: The reinvention of nature*. New York: Routledge.

Harris, L. 2009. Gender and emergent water governance: comparative overview of neoliberalized natures and gender dimensions of privatization, devolution and marketization. *Gender, Place & Culture*. 16 (4):387–408. doi: 10.1080/09663690903003918.

Harris, L. 2014. Imaginative geographies of green: difference, postcoloniality, and affect in environmental narratives in contemporary Turkey. *Annals of the Association of American Geographers*. 104 (4):801–815.

Holling, C.S., L.H. Gunderson, and G.D. Peterson. 2002. Sustainability and panarchies. In *Panarchy: Understanding transformations in human and natural system*, edited by L.H. Gunderson and C.S. Holling, 63–102. Washington, D.C.: Island Press.

Hovorka, A.J. 2006. The no. 1 ladies' poultry farm: a feminist political ecology of urban agriculture in Botswana. *Gender, Place & Culture*. 13 (3):207–225. doi: 10.1080/09663690600700956.

Jarosz, L. 2004. Political ecology as ethical practice. *Political Geography*. 23 (7):917–927. doi: 10.1016/j.polgeo.2004.05.014.

——. 2011. Nourishing women: toward a feminist political ecology of community supported agriculture in the United States. *Gender, Place & Culture*. 18 (3):307–326. doi: 10.1080/0966369x.2011.565871.

Keil, R. 2003. Urban political ecology. *Urban Geography*. 24 (8):723–738.

Kimura, A.H. and Y. Katano. 2014. Farming after the Fukushima accident: a feminist political ecology analysis of organic agriculture. *Journal of Rural Studies* 34:108–116.

Lahiri-Dutt, K. 2006. *Fluid bonds: Views on gender and water*. Kolkata: Stree.

Leach, M. 2007. Earth mother myths and other ecofeminist fables: how a strategic notion rose and fell. *Development and Change*. 28 (1):67–85.

Lynch, B.D. 1991. Women and irrigation in highland Peru. *Society & Natural Resources*. 4 (1):37–52.

McLeman, R.A. and L.M. Hunter. 2010. Migration in the context of vulnerability and adaptation to climate change: insights from analogues. *Wiley Interdisciplinary Reviews: Climate Change*. 1 (3):450–461.

Mollett, S. and C. Faria. 2013. Messing with gender in feminist political ecology. *Geoforum*. 45:116–125. doi: 10.1016/j.geoforum.2012.10.009.

Moss, P. 2002. Taking on, thinking about, and doing feminist research in geography. In *Feminist geography in practice: Research and methods*, edited by P. Moss, 1–20. Oxford: Blackwell.

Mountz, A. and J. Hyndman. 2006. Feminist approaches to the global intimate. *Women's Studies Quarterly*. 34 (1/2):446–463.

Nash, J. 2005. Women in between: globalization and the new enlightenment. *Signs*. 31 (1):145–167.

Nelson, V. and T. Stathers. 2009. Resilience, power, culture, and climate: A case study from semi-arid Tanzania, and new research directions. In *Climate change and gender justice*, edited by G. Terry, 57–70. Oxford: Oxfam, Great Britain and Rugby, UK: Practical Action Publishing.

Nielsen, S.T. 2010. Coastal livelihoods and climate change. In *Reducing poverty, protecting livelihoods, and building assets in a changing climate: Social implications of climate change in Latin America and the Caribbean*, edited by D. Verner, 123–165. Washington, D.C.: The World Bank.

Nightingale, A.J. 2003. A feminist in the forest: situated knowledges and mixing methods in natural resource management. *ACME: An International E-Journal for Critical Geographies*. 2 (1):77–90.

——. 2011. Bounding difference: intersectionality and the material production of gender, caste, class and environment in Nepal. *Geoforum*. 42 (2):153–162. doi: 10.1016/j.geoforum.2010.03.004.

O'Reilly, K. 2006. "Traditional" women, "modern" water: linking gender and commodification in Rajasthan, India. *Geoforum*. 37 (6):958–972. doi: 10.1016/j.geoforum.2006.05.008.

Ogden, L., N. Heynen, U. Oslender, P. West, K.-A. Kassam, and P. Robbins. 2013. Global assemblages, resilience, and earth stewardship in the anthropocene. *Frontiers in Ecology and the Environment*. 11 (7):341–347.

Peet, R. and E. Hartwick. 2009. *Theories of development: Contentions, arguments, alternatives*. Second ed. New York: Guilford Press.

Pickett, S.T.A., M.L. Cadenasso, and B. McGrath. 2013. *Resilience in ecology and urban design: Linking theory and practice for sustainable cities*. New York: Springer.

Postel, S. 2014a. Groundwater depletion in Colorado River basin poses big risk to water security. *Water Currents*, http://newswatch.nationalgeographic.com/2014/07/30/groundwater-depletion-in-colorado-river-basin-adds-big-risks-to-water-security/. Accessed August 1, 2014.

——. 2014b. A sacred reunion: The Colorado River returns to the sea. *Water Currents*, http://newswatch.nationalgeographic.com/2014/05/19/a-sacred-reunion-the-colorado-river-returns-to-the-sea/. Accessed August 1, 2014.

Radel, C. 2009. Natures, gendered. In *International encyclopedia of human geography*, edited by R. Kitchin and N. Thrift, 331–336. London and New York: Elsevier.

——. 2012. Gendered livelihoods and the politics of socioenvironmental identity: women's participation in conservation projects in Calakmul, Mexico. *Gender, Place & Culture*. 19 (1):61–82.

Resurrección, B.P. 2013. Persistent women and environment linkages in climate change and sustainable development agendas. *Women's Studies International Forum*. 40:33–43. doi: 10.1016/j.wsif.2013.03.011.

Robbins, P. 2004. *Political ecology: A critical introduction*. Malden, MA: Wiley-Blackwell.

Rocheleau, D. 1995. Maps, numbers, text, and context: mixing methods in feminist political ecology. *The Professional Geographer*. 47 (4):458–466. doi: 10.1111/j.0033-0124.1995.458_h.x.

Rocheleau, D. and R. Roth. 2007. Rooted networks, relational webs and powers of connection: rethinking human and political ecologies. *Geoforum*. 38 (3):433–437.

Rocheleau, D., B. Thomas-Slayter, and E. Wangari (eds). 1996. *Feminist political ecology: Global issues and local experiences*. New York: Routledge.

Romero-Lankao, P. et al. 2014. North America. Chapter 26. IPCC Working Group II Fifth Assessment Report. http://ipcc-wg2.gov/AR5/images/uploads/WGIIAR5-Chap26_FGDall.pdf. Accessed April 22, 2014.

Salleh, A. 2009a. *Eco-sufficiency and global justice: Women write political ecology*. London: Pluto Press.

——. 2009b. Ecological debt: Embodied debt. In *Eco-sufficiency and global justice: Women write political ecology*, edited by A. Salleh, 1–14. London: Pluto Press.

Schmidt, L.A. and S. Buechler. 2014. Honduran boys confronting adversity: urban multi-locality and kin mobilization. *Children's Geographies*. 12(2). doi: 10.1080/14733285.2014.890390.

Stargardter, G. and G. Palencia. 2014. After U.S. deportation, a Honduran mother and daughter's uncertain fate. Tegucigalpa, Honduras. Reuters. July 18, 2014. www.reuters.com/article/2014/07/18/us-usa-immigration-deport-insight-idUSKBN0FN2CD20140718. Accessed. July 19, 2014.

Sultana, F. 2009. Fluid lives: subjectivities, gender and water in rural Bangladesh. *Gender, Place & Culture*. 16 (4):427–444. doi: 10.1080/09663690903003942.

—— and A. Loftus. 2012. *The right to water: Politics, governance and social struggles*. Earthscan water text. London: Routledge.

Sundberg, J. 2008. Placing race in environmental justice research in Latin America. *Society & Natural Resources*. 21 (7):569–582.

Truelove, Y. 2011. (Re-)conceptualizing water inequality in Delhi, India through a feminist political ecology framework. *Geoforum*. 42 (2):143–152. doi: 10.1016/j.geoforum.2011.01.004.

Vayda, A.P. and B.B. Walters. 1999. Against political ecology. *Human Ecology*. 27 (1):167–179.

Walker, B. and D. Salt. 2006. *Resilience thinking: Sustaining ecosystems and people in a changing world*. Washington, D.C.: Island Press.

Walker, P.A. 2005. Political ecology: where is the ecology? *Progress in Human Geography*. 29 (1):73–82.

Waterman, J. 2012. "Where the Colorado Runs Dry." *The New York Times*, February 14.

Watts, M. 2000. Political ecology. In *A companion to economic geography*, edited by T. Barnes and E. Sheppard, 257–275. Oxford: Blackwell.

Zimmerer, K.S. and T.J. Basset. 2003. *Political ecology: An integrative approach to geography and environment-development studies*. New York: Guilford Press.

Zwarteveen, M. 1997. Water: from basic need to commodity: a discussion on gender and water rights in the context of irrigation. *World Development*. 25 (8):1335–1349.

Zwarteveen, M., A. Ahmed, and S. Rimal Gautam (eds). 2012. *Diverting the flow: Gender equity and water in South Asia*. New Delhi: Zubaan books.

Part I

Feminist political ecology and large-scale water resource management

2 Interrogating large-scale development and inequality in Lesotho

Bridging feminist political ecology, intersectionality, and environmental justice frameworks

Yvonne A. Braun

Large dams serve as the quintessential example of a utilitarian and market approach to the environment, valuing rivers and river basins exclusively with technical and market-based criteria. While dams are being decommissioned in the United States, in some cases stemming from the negative impacts of dams on riverine habitats and ecological consequences, dam building continues to be promoted in developing countries with estimates of 40–80 million people affected by large dams worldwide (World Commission on Dams 2000). The promises of large dams are narrated globally through the shared language of development and nationalism; these interests are particularly salient to developing states, international financial institutions and elites in the large-scale infrastructure industry. In Lesotho, the commodification of water and the building of large dams sit at the center of an extractive, export-oriented strategy to generate national revenues by selling water to South Africa. These projects are, as critics feared, affecting tens of thousands of Basotho people during the anticipated 30-year period (1987–2017) of dam construction and likely well beyond dam completion.

Consequences of the Lesotho Highlands Water Project (LHWP) include remote highlands families' losses of agricultural fields, grazing lands, and communal resources. Perhaps most ironically, access to clean and reliable water has been compromised by the conversion of rivers into reserves of marketable water. Women, traditionally charged with fetching water for their households, have reported walking up to 2.5 miles more per round-trip to access water (Braun 2005b) since the building of the LHWP's Katse Dam, one of five dams in this 30-year project.

In this chapter, I argue for the utility of bridging feminist political ecology with insights from intersectionality and environmental justice frameworks for contemporary analysis of macro projects like the LHWP's US$8 billion water scheme. These combined analytical approaches make visible the lived realities of the people and the environment that are traditionally rendered invisible in dominant ideologies, such as market and technocratic visions of development through dams. These realities are furthermore rendered invisible in the physical landscape, through the actual transformation of river systems and the dislocation of communities.

I explore these linked frameworks by analyzing remote, rural communities' negotiation of the challenges posed by their rapidly changing environment, especially water resources (due to LHWP and large-scale development). Water is crucial to the sociopolitical and economic stability of the region; this chapter, therefore, aims to demonstrate how a feminist political ecology framework, sharpened by insights from environmental justice and intersectionality, can deepen our understanding of the lived complexities and questions of justice raised by large-scale water development.

Large-scale dams and infrastructure projects in the Global South

Large-scale dam and infrastructure projects remain a common and controversial means toward development and poverty reduction in the Global South (Scudder 2006; World Commission on Dams 2000). The social and environmental consequences of large dams have been extensively debated, critics arguing the local effects—resettlement, dislocation, decreased access to natural resources—reveal the contradictions and flawed logic of dams as national development solutions (Scudder 2006). Typically the poor and marginalized absorb a disproportionate share of the costs of these projects (Roy 2000), such as development-induced resettlement (Hussain 2008), leading to local and global opposition as with the Narmada Projects in India (Khagram 2004), and long-term consequences such as higher rates of poverty for successive generations as the Gwembe Tonga who were relocated for Zambia's Kariba Dam (Scudder 2006) have experienced. The gendered, raced, and classed nature of the institutional priorities, processes, and practices, as well as the consequences of globalized dam building, have received less sustained attention.

As the significant social and environmental consequences of dams are justified through national development, the communities that absorb the most direct consequences of large-scale development become part of a rationale of "local pain for national gain" or of what is best for "the greater common good" (Roy 2000, 58, 43). Large-scale dam projects are almost without exception sited in poor, rural, and politically, ethnically, or socially marginalized communities, ironically structuring the "local pain" to disproportionately affect already disempowered populations. I argue these patterns associated with large-scale development, such as the building of mega-dams, are best interpreted through the combined fields of environmental justice, feminist political ecology, and intersectionality.

Environmental justice

In the 1980s, the concept of environmental justice emerged in the United States as a way of describing the patterns of disproportionate burdening of poor, low-income communities, and communities of color with the wastes of modern society—pollution, toxins, and locally unwanted land uses (LULUs) (Bullard 1990). As environmental justice is increasingly globalized, the way that it is defined, interpreted, and employed varies within social, political, and cultural contexts and histories (Walker 2012).

Bannerjee and Bell argue that, as environmental justice broadened the initial concerns of environmental racism, the evolution of environmental justice incorporates the concept of environmental democracy, opening the discussion to consider more directly the "range of inequalities, including questions of the fair distribution of environmental goods, not only the fair distribution of environmental bads" (2008, 444–5).

Freshwater is, perhaps, the quintessential environmental good—a life-sustaining resource necessary for survival, integral to well-being and livelihoods in the rural agriculturally based communities of the Lesotho highlands. Large dams fundamentally alter riverine systems and water resources, creating rapid environmental change in the communities where they are situated. Where large dams form part of an extractive-oriented development plan, such as in the LHWP, water becomes commodified and re-organized in a logic of utilitarian and market-based development solutions. Local communities are generally not included in the decision-making about large dam placement and construction, yet they largely bear the burdens of the changes, such as reduced or changed access to water and riverine resources, and often in gendered ways.

Feminist political ecology and intersectionality

Political ecology, a multi-disciplinary approach that addresses concerns of both political economy and cultural ecology, emphasizes how environmental resources and decision-making are material and political, and are affected by different interests, investments, and power (Peet and Watts 2004). Feminist political ecology has long enriched the field of political ecology by bringing critical attention to the role of gender in the structuring of relations to resources and in the production of knowledge (Rocheleau et al. 1996). Gendered analysis of resource struggles, ecological practices, and power relations within relevant historical, socio-cultural, political, and geographical localities revealed more complex and critical understandings within human-environment and nature-society relations.

Recent scholarship in feminist political ecology has begun to explicitly engage developments in feminist theory more broadly, highlighting the synergies between contemporary theorizing and analysis of intersectionality and "how human-environment relations intersect with various social relationships, such as class, ethnicity, and gender" (Radel 2009, 335). Contemporary feminist scholarship increasingly centralizes intersectionality, widely understood as the theorization and analysis of intersecting and mutually constituting axes of inequality and experience, such as race, class, and gender (Crenshaw 1991). Intersectional analysis (McCall 2005) reveals how social structures, relations, processes, and policies shape and constrain social experiences along axes of inequality bringing attention to multiple, mutually constituted identities and social locations.

While the intersectionality approach advances efforts to capture the complexities of the lived social life of individual and group identities, institutions, structures, and systems across historical, geographical, and cultural contexts, the environment has largely gone unconsidered. Feminist political ecology brings attention to the environment, revealing how resource struggles and their consequences are gendered, how environment and gender are co-constructed, and "how lived experiences and practices are productive of, and produced through, gendered ideologies, structural power relations, and processes of both local and global change" (Truelove 2011, 145). By integrating intersectionality more intentionally, feminist political ecology considers how environmental resources and struggles are sites for the contestation and reproduction of social differences, such as gender, race, and class, and, as discussed above, as struggles over environmental justice.

By engaging insights from the environmental justice literature and feminist theorizing on intersectionality within feminist political ecology's attention to political, cultural, and geographical contexts, this chapter attempts to show the value of conversing across these interdisciplinary fields and concepts. Such analyses enable feminist political ecologists to explain and illuminate the complex lived experiences of rapidly changing water landscapes as a result of large-scale development, such as large dams, and the ways that large dam projects may be best understood as struggles with environmental justice.

Research context

Lesotho is a small, mountainous, landlocked country surrounded by South Africa on all borders (see Figure 2.1). With a population of approximately two million people, over 80 percent of the Basotho population lives in rural areas using gendered livelihood strategies that rely on agriculture, livestock farming, wage labor, and informal economic activities (Epprecht 2000). Historically, Lesotho conforms to a classic model of a dual economy, with a primary subsistence sector in the rural regions of the country and a secondary migratory labor reserve economy dependent on South Africa's mining and industrial sector. Male migration[1] to South African mines peaked in the late 1970s and 1980s when up to 35 percent of active wage-earning men from Lesotho were working in South Africa (Central Intelligence Agency 2010), often making women *de facto* heads of household.

As a traditionally patrilineal and patrilocal society, most households and inheritances are organized around the lineage of men (Epprecht 2000) and, until recently, the customary and legal policies in Lesotho marked women and girls as legal minors throughout their adult lives. Women of any age needed fathers, and subsequently husbands and sons, to access the full rights of adulthood in Lesotho—land ownership, bank accounts, loans, LHWP compensation, and so

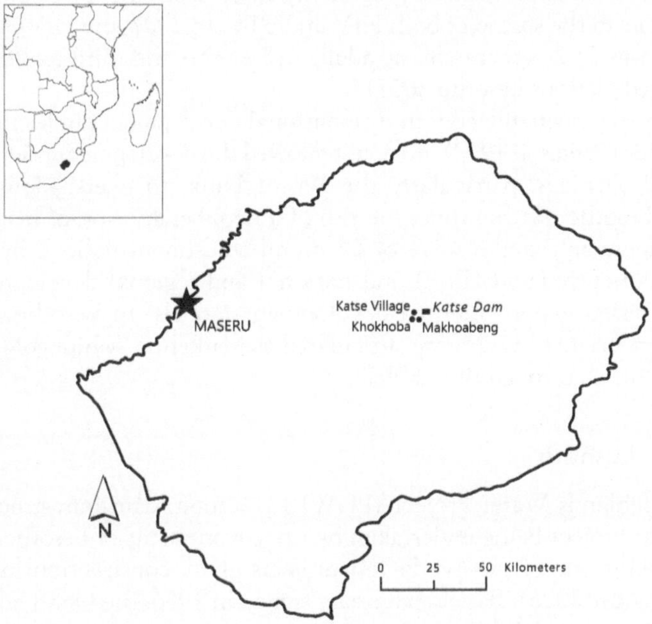

Figure 2.1 Map of Lesotho with location within Southern Africa in insert
Source: Map created by Leslie A. McLees

on (Braun 2005b).[2] As a result, money has long been a domain of contestation within Lesotho households, and the gendered struggle to gain access to resources is a dynamic, ongoing process (Letuka et al. 1997; Murray 1981).

Within this gendered economy, women are often the primary farmers and responsible for the social reproduction of the household, including food preparation, cooking, collecting water and fuel, and childcare, whereas men are involved in wage labor and some aspects of farming and raising livestock. With the dramatic retrenchment of male miners since the early 1990s (Global Policy Network 2006), migration for work in the South African mines has decreasingly been an option for men from poorer, rural areas such as those affected by the LHWP. Local opportunities for wage labor in these extremely remote highlands areas remain very limited, and most wage labor still involves migration to South Africa or to urban areas in Lesotho. In the last decade, it is increasingly women who are migrating to urban areas to find work in the highly competitive domestic service field and the textile and garment industry. Estimates indicate that up to 50,000 Basotho were employed in textile and garment factories during the peak of employment in the early 2000s but jobs in these foreign-owned factories have greatly diminished since 2004 (Bennett 2006). These historical and contemporary migration patterns, a direct result of persistently high rates of poverty and unemployment[3]—approximately 60 percent[4] (World Bank 2010) and 30–40 percent over the last two decades, respectively (United Nations Lesotho 2007)—as well as structural relations of inequality regionally and globally, have also proven to be major factors in the spread of both HIV and STIs. In 2007, the HIV/AIDS prevalence rate was 23.2 percent among adults in Lesotho, the third highest in the world (United Nations Lesotho 2007).

Lesotho, deemed perpetually poor in international development circles in the early to mid-1980s (Ferguson 1990), no doubt allowed the development priorities of international agencies, particularly the World Bank, to overshadow the government of Lesotho. At the time, the rise of a neoliberal vision of markets and of development initiated a wave of structural adjustment policies by the International Monetary Fund (IMF) and national and regional development agendas that stressed export-oriented development models. In Lesotho, this neoliberal, export-oriented vision was articulated as marketing "white gold," or water, to South Africa through the LHWP.

Large dams in Lesotho

The Lesotho Highlands Water Project (LHWP) is a multi-dam infrastructure and development project being undertaken by the governments of Lesotho and South Africa and is one of the five largest projects under construction in the world today (Scudder 2006). Based on a treaty signed in 1986, the US$8 billion project is funded in part by the World Bank, the African Development Bank, the European Community, and several European funding agencies, and is being implemented by the parastatal Lesotho Highlands Development Authority (LHDA) in Lesotho. The water delivery scheme includes five dams linked to cross-national tunnels planned to be constructed in four phases over 30 years (1987–2017) and will sell and deliver water from rural Lesotho to the industrial region of South Africa (LHDA 1986).[5] Currently three dams have been completed: Katse, 'Muela, and Mohale.

Losses from and consequences of the LHWP affected an extensive geographical area in Lesotho, with concentrated effects in communities around the dams and reservoirs (see Figure 2.2). Households reported losing important material, social, cultural, and environmental resources such as agricultural, grazing, and garden lands; access to water sources for drinking, household, and agricultural use; forests for fuel wood; river basins for soils, rocks, and grasses used in constructing houses and cultural displays, such as painting designs on houses; and wild herbs and vegetables for food and medicinal purposes. Consistent with the social and environmental consequences of other large dams (Scudder 2006), these losses generally burdened already poor Basotho households and created especially intense pressures on women (Braun 2005a, 2005b). The highlands areas chosen as dam construction sites contain some of the most remote communities within Lesotho, with some of the highest rates of unemployment and destitution (Sechaba Consultants 1994; Tshabalala and Turner 1989). At the beginning of the LHWP, 60 percent of households in Katse and 'Muela fell below the average income for each area and were considered "very poor" (Panel of Environmental Experts 1991, 25; Tshabalala and Turner 1989, 9).

Figure 2.2 Village of Makhoabeng near Katse Dam
Photo credit: Yvonne A. Braun

While material losses resulting from the construction of the dams themselves were the most obvious impact, households were affected in multiple ways. Approximately 20,500 residents in over 120 villages were affected in the Phase 1A area (Tshabalala and Turner 1989), including the vast infrastructure constructed for the dam and tunnel system through Lesotho to South Africa, the construction of roads through mountainous areas, and the construction of employee "camps" for foreign workers at the dam sites.

As affected communities negotiated the influx of foreign and regional Basotho[6] migrants to the dam areas, the confluence of retrenchment from South African mines and losses of agricultural fields—a significant means of production—to the LHWP added considerable economic pressure on rural highlands households to find alternate means. In this context, the promises of employment with the LHWP and the hopes of a better life through large-scale development were especially important to local residents.

In the case below, I attend to the questions residents raised about decision-making and democracy and how they expressed critiques of large-scale development through the lens of their experiences in these embodied landscapes. I demonstrate how employing a feminist political ecology approach to seeing large-scale development, such as the building of mega-dams, as environmental (in)justice opens up the possibility to render visible the lived experiences of people directly affected by the LHWP within a broader analysis of just development.

Methods

Ethnographic fieldwork typically involves multiple forms of data collection and participation in the research setting including, but not limited to, participant observation coupled with informal and formal interviews (Berg 2009). Participant observation is a method of observing social phenomena while being part of a research setting, in this case living in communities directly affected by the LHWP and participating in the daily routines of life in a household and village. Informal interviews are conversations or discussions that take place in the course of doing research and that offer the ethnographer ongoing opportunities to gather information and understanding from participants over time. In contrast, formal interviews are guided conversations arranged and led by the researcher with varying levels of structure. Experiences and data gathered through participant observation and through formal and informal interviews allow the researcher opportunities for deeper engagement in and greater understanding of the research setting and the topic of inquiry.

This chapter draws from primary multi-site ethnographic fieldwork conducted in Lesotho in 1997 and 2000–2002 in all three active dam areas. Both research periods included participant observation, formal and informal interviews, and documentary research.[7] I also volunteered and collaborated with local non-governmental organizations, attended relevant meetings and conferences, and interviewed development consultants and officials during my fieldwork.

Living in the communities directly affected by the LHWP, I conducted countless informal interviews with people living proximate to the project as part of our everyday interactions as neighbors and acquaintances, and as friends and "family." Informal interviews were recorded in field notes. Formal interviews were conducted in a range of villages[8] affected by the LHWP and were semi-structured and open-ended, designed to ask participants a similar set of questions while affording them the opportunity to elaborate and introduce new topics and ideas. Most interviews were conducted in the highlands areas, with a few in resettlement sites in urban areas, and ranged from 15 minutes to one-and-a-half hours. Interviews were conducted in both Sesotho and English,[9] were recorded when possible, then transcribed and translated into English as needed. Formal interviews are summarized by dam location and fieldwork period in Table 2.1.

The villages in these very remote and rural highlands areas are some of the poorest in Lesotho, one of the poorest countries in the world. In 1996, when

Table 2.1 List of formal interviews by dam location, 1997 and 2000–2002

Location	*1997*	*2000–2002*	*Total*
Katse	16	94	110
'Muela	13	76	89
Mohale	0	93	93
Total formal household interviews	29	263	292

Phase 1A was in construction, the average yearly income for a Basotho household was US$440 (Pottinger 1996) and was likely much lower in the highlands areas. Over 50 percent of the total population currently lives below the international poverty line of less than US$1.25 a day.

These national statistics give a sense of conditions at the village level, although village level data are not available. Ethnographic data, reinforced by national and international information, suggest that while conditions within villages vary, remote highlands communities in the Katse Dam area generally share in common extreme poverty, the absence of modern social infrastructure such as electricity, running water, and plumbing, and have limited paved roads. I lived in in the village of Khokhoba, also known as Ha Mensel, for extended periods during both research trips due to its proximity to the Katse Dam and related facilities. Indeed, most of these facilities were built on lands the development authorities appropriated from residents of Khokhoba and, to a lesser extent, its neighboring village of Makhoabeng. My daily experiences in Khokhoba allowed me to witness the consequences of the LHWP in the lives of community members and to have countless informal discussions with local residents regarding their perceptions and experiences with the LHWP.

Like other villages in the area, Khokhoba and Makhoabeng are largely organized around a subsistence-oriented agricultural economy paired with migratory wage labor employment, reflecting the expectations from existing literature about adult men's and women's roles in labor markets across the region. In my formal interviews, one-third of respondents reported no formal education while four participants (1.5 percent) reported higher education, with one of them living in the Katse area. Participants in the Katse area overwhelmingly declared farming, mining, and looking for work as men's primary occupations and household work and farming as women's primary occupations, paralleling what is reflected in the literature. Reflecting on the very limited opportunities available for formal employment in the highlands area, interviewees reported casual labor or piece jobs, looking for work, or none as the secondary occupations for themselves and other adults in their households and declared their hopes, often unfulfilled, for work with the LHWP.

As I narrate the analysis below, I draw on this primary data—fieldwork, observations, and informal and formal interviews—using themes derived from my data and subsequent coding and analysis. All interviewees' names are pseudonyms.

Large-scale development and dams in Lesotho

> The history of large dams and affected communities in Southern Africa has been one of broken promises and incalculable losses…We have been forced to move against our will without knowing when or where we would be going, and without a way for our concerns or objections to be heard. We have not been treated with dignity, nor with respect for our customs, our ancestors or our children. We have shouldered the burden of large dams, but we have

> enjoyed very few of the benefits. In short, large dams have been devastating to many of our communities.
>
> (Southern African hearings for communities affected by large dams 1999)

National strategies for economic development commonly employ infrastructure projects such as mega-dams, hydroelectric power, and, as my case captures, international trade in water or "white gold" as it is known in Lesotho. As the remote, rural poor in Lesotho negotiate the impacts of the LHWP in their lives, they experience the intrusion and privileging of the transnational and local interests that organize and finance projects such as the LHWP. As some lose their land and homes in this process, many more also experience a changed landscape that includes the loss of access to significant natural and cultural resources such as land, water, and plants, the loss of the ways they have lived their lives, and the loss of their pre-dam livelihoods.

Embodied landscapes of development

The LHWP mega-project became the symbol of Lesotho's long-term national development plans for economic growth and human development and, according to the economistic visions of World Bank consultants, the best path to the much-needed promise of neoliberal development in Lesotho (Braun and Dreiling 2010). Because neoliberalism promotes an extension of markets, the valorization of property rights, and a privatization of public resources, political rights are diminished while market rights are elevated. In this neoliberal vision of development, any sense of collective justice is a misnomer, as individuals are compensated through quantifiable values via the technocratic authority of the development agency and the state, and individuals are cut loose from land as detached producers-consumers. All qualitative, relational, and communal values are rendered valueless and invisible. As with the river ecosystems, the village life in the remote highlands faced irreparable change under the guise of exporting white gold for the good of the nation. With the emphasis on GDP growth and large-scale infrastructure development to export water to South Africa, Lesotho's military government in the 1980s fused nationalistic and traditional patriarchal idioms with neoliberalism. For local landowners in the vicinity of the LHWP, their lived experience of development often entailed displacement and dispossession of their homes and lands (Braun 2011a), integral parts of residents' rural livelihoods and the fabric of their shared history.

In order to implement the LHWP in these extremely remote and poor highlands communities, expertise from other countries was brought in and housed in long-term facilities proximate to affected local communities. Development authorities constructed a modern, gated housing community for foreign workers brought in to work the best-paid jobs at the first and largest site, Katse Dam (Braun 2011b). The contrast between the residential, recreational, and job benefits the foreign workers had access to and the conditions in which local Basotho lived make manifest an intentional physical segregation and the creation of unequal space (Braun and McLees 2012).

While thousands of people lost their land and subsistence livelihoods to the LHWP and suffered increased dependence on the project for economic support, the available work opportunities were circumscribed by technical and organizational priorities made by the development institutions. These limited, and often precarious, opportunities run counter to the promises of development[10] propagated in the neoliberal and modernist discourses, as well as the specific ways that the LHWP was sold to Basotho people based on these ideals. The promises included neoliberal notions that all Basotho would benefit from the national income generated by the mega-project, purportedly as a result of the subsequent socio-economic development that the LHWP would bring. Consider this quote from a development authority promotional brochure: "The direct investment and increased economic activity created by the project will generate a significant source of income for Lesotho's long term development. Perhaps more importantly will be the creation of much needed employment opportunities" (quoted in Pottinger 1996, 5). The employment opportunities promised were particularly convincing to the rural poor struggling to manage rising retrenchment from the mines, increasing agricultural competition and declining productivity, and widespread high rates of unemployment.

The foreign worker housing, of course, was built on the grazing lands taken from the adjacent village, Khokhoba, as the dam and its supporting infrastructure submerged homes, lands, and significant riverine resources. Development authorities mitigated these losses using a system of compensation fraught with complications and challenges, resulting in grievances filed by people directly affected and local organizations working on their behalf.

Privatizing the commons

Communities, households, and individuals living proximate to the LHWP also lost or endured reduced access to natural resources not owned privately, or common resources. For rural households in these remote communities, these resources were integral to the means of production and reproduction of the household and norms of commons' usufruct governed access to and shared responsibility toward these commons. For example, local river basins provided water for multiple purposes, stones, soils, grasses, wood, and fertile farmlands. People understood the value of these commons for themselves and their community. Reduced or no access to these resources represented significant losses to rural households that have relied on these ecosystems for generations. 'Mathebiso Mpiti, a sharecropper and mother of two, whose landless household relied on cultivating others' fields and utilizing common resources said:

> [Our life] has changed because most of our resources—we used to find them at the river here. So, as the dam has covered things, grass and others are no

> longer available. We have no land to cultivate and no food to gather. We have nothing, it is all in that dam now.

Similar to agricultural fields, these common resources also have productive and generative potential for households that use their local knowledge, often gendered, to cultivate and transform these resources into income—particularly in an environment with limited employment opportunities. 'Malefetsane Nthako, an elderly grandmother from the village of Ha Mapeleng near Katse Dam, collected *loli*, a type of grass, in the river basin in order to make woven goods that she then sold in the lowlands areas. The reservoir at Katse submerged these grassy areas, leaving 'Malefetsane and others like her with great uncertainty as to how to replace the income this resource provided:

> We would get *loli* [from the river], making baskets and sifters. Now it has closed [the potential to make woven goods]. We knew that when we made them, we could go to the lowlands and they would buy them [and] we would make a living . . . so now we don't know what we [will] do . . . [to] make a living.

'Mathabane Likong, a farmer and primary caregiver to her two grandchildren, suggested the resources of the river had not only economic value to households but reflected and constituted important social and cultural values:

> Down there at the dam there are herbs that I used to dig up from there; now that place is covered. I no more get those medicines—I mean a particular herb will be found at a particular place, they don't just grow. I mean they chose where to grow. I no more get them. We collected wood. We dug up soils and we painted like the Basotho people . . . those soils are not there now, they are finished . . . the dam has covered up those soils.

Common resources in these river basins, so integral to these rural communities, were largely seen as underutilized in the development authorities' economistic evaluations prior to the LHWP. As river basins rich with these goods were flooded as part of the LHWP, communities and households were not compensated for these non-privately owned resources despite their regular use and local value. The loss of these resources, particularly without compensation, proved significant for economic and cultural reasons, and the effects were gendered. Indeed, poorer households' reliance on these gathered resources as "low-cost" or "free" goods had been made possible by women's uncompensated labor, who spent time and effort learning about different plants and soils, as well as the labor of locating, gathering, and using these resources.

As the LHWP flooded the most accessible and best sources of these natural resources, the rapid environmental changes intersected with existing gender ideologies to create burdens on women to devote more of their labor to finding replacement resources within this changing landscape. This uncompensated, gendered labor adds value to the household's way of life both economically and culturally, but is largely taken for granted as not "real work" or invisible in terms of the dominant discussions about the lived consequences of large-scale river basin development.

Resettlement of households and their homes as a result of the LHWP also initiated significant environmental and social changes in the lives of those resettled. The experiences of communities varied in terms of how and where they were resettled, with some being relocated together, others moved to existing communities (called "hosts"), some moved within the general vicinity while others moved to distant places. While it is not possible to discuss the full range of these experiences here, there were patterns to resettlement that included most people not receiving fields, but rather garden space, and new houses being constructed of concrete and metal, as opposed to traditional mud, stone, and thatch. The changes forced by resettlement created a great deal of uncertainty, particularly as the loss of means of production (fields) and the change of environment meant a loss of common resources on which they had depended. As 'Malimakatso Moubane, a resettled farmer and mother of four, described, survival in these relocation areas demanded money at a time when these households were least equipped to have access to money:

> The problems we meet [are] to lack something to make fire with, we do not have food, and—and you know in the mountains there were lots of food lying around, we would go to the veld and collect vegetables, we cultivated, we survived by food that we grew. Now, here, when you have got no money, everything is money.

Households who were resettled in peri-urban or urban areas confronted similar struggles of a rapidly changed economic and environmental landscape. While each new house had the potential for its own outside water tap, resettlers to these areas were surprised to find the potential ease of water access was made impossible due to the high cost of water and connecting to the utility. Many households could not afford to purchase the water needed to meet their daily needs for drinking, cooking, bathing, and cleaning directly. Relative class privilege positioned some families who could afford to be able to subsidize their own water bills by selling water by the bucket to neighbors "off the grid." Gender ideologies still positioned women in this context to largely be charged with the gathering of water and the household tasks associated with it. Women creatively managed high prices for water and water scarcity by sharing water for tasks (such as laundry), exchanging services for water, and taking turns or pooling limited resources with other women to purchase water.

Tragically, the greater reliance on money also affected the ability of those resettled in a new environment to take care of their health with scarce financial and environmental resources. The ecological knowledge that many rural highlanders, particularly herbalists, had gained over generations was often rendered useless after relocation as demonstrated in the words of Vincent

Kolisang, a herbalist, farmer, and father of eight: "The medicines we were used to digging up ourselves, to make ourselves feel better, are not available here . . . we live by going to the chemist [pharmacy] to buy medicines."

It was not only those forcefully relocated who suffered changes in relation to their environment as a result of the LHWP. Perhaps most ironically, many remote highlands families who continued to live proximate to the mega-dams lost access to clean, reliable water sources as the easily accessible rivers were transformed into massive reservoirs of water to be sold. Changes from damming the riverine system caused a number of consequences, including altering a number of the natural springs that supplied clean, fresh water to local communities. 'Mathato Lejone expressed frustration about losing the water source she relied on for years explaining, "This dam has brought hardship. Now our springs have vanished. These were near to us, but now we must go far to get water." 'Mathato's experience was a common one as many people reported springs drying up around the Katse basin after the creation of the new reservoir and women talked about having to find new sources of water, sometimes walking significantly further to do so (see Braun 2005b).

Local people also experienced new forms of regulation that limited their access to water, as the once free flowing resource became commodified and owned under new policy doctrines designed to generate a specific model of economic development. People's narratives about water revealed these policy contradictions in their daily lives, often in gendered ways: women walking further to find water sources to meet the everyday needs of their families; some taking the risk of "stealing" water from the reservoirs, now a criminal offense; others loathing the reservoirs that created an impasse between themselves and relatives; and still others describing the reservoirs as a site of fear and desperation, a place where some men drowned themselves when hardships resulting from the LHWP became too much.

These seemingly individual narrations, or the policy stories people tell (Yanow 1995), weave together a much larger tapestry that reveals the contradictions of national development strategies in this struggle over environmental justice. Quotes presented below from two elderly people who reflected on their lived experience through broader critiques of development reveal some of these contradictions and the resulting struggles. The first quote is from a female farmer from Ha Mapeleng in the Katse area: "We are leading a hard life . . . coldness is very much. Since the dam there is no summer, there is no winter. It is very cold. We never put [heavy clothes] down. I can't see where development is." And this strong statement from a male farmer and chief of a village resettled due to the Mohale Dam:

> I don't think it can be beneficial [to us], even if the LHWP has benefited Lesotho. But now the advantage [of the LHWP] is hurting us, the Basotho here, since we don't get our rightful compensations. . . . So we are supposed to survive, so the LHWP benefits. As for us who are affected we don't benefit, we are wrongly treated.

Discussion and conclusion

As the development authorities and the state appropriated the material and natural resources of local communities for national economic growth, they codified the value of these resources in ways that reflected a privileging of private property, men's ownership of resources, and formal, wage work. The project authorities' limited interpretation of the value of agricultural production in these rural communities minimized the productive and generative nature of agricultural fields for household consumption, income, and entrepreneurship as demonstrated above.

In particular, the work of rural people, their labor and their businesses, were undervalued or unvalued as livelihoods and left out when calculating the costs of the consequences of social and environmental change involved in large-scale dam development. Even as rural highlanders directly relied on natural resources to fulfill their daily responsibilities—as farmers, cultivators of gardens and riverine forests, gathering fuel and stones, collecting wild vegetables and herbs, and fetching water—their labor was naturalized and unacknowledged, undervalued to the extent that it did not warrant compensation in the calculations made by development authorities. The exploitation of natural resources and ensuing environmental change were presented as development in this context, which was complexly interwoven and supported by modernist assumptions about the invisibility and lack of productivity of rural people's livelihoods.

When interpreted through an environmental justice framework, these evaluative practices are an extension of an ideological position rooted in neoliberal modernization. On the one hand, large-scale, utilitarian development solutions such as large dams are seen as efficient solutions for using underutilized natural resources such as rivers and modernizing "underdeveloped" or poor societies, the consequences being justified through an argument of national development for "a greater good" (Roy 2000).

Environmental justice and feminist political ecology, however, raise questions about which communities absorb the otherwise hidden or invisible consequences and who receives disproportionate benefits of "the greater good." In this case, the disruption of rural communities—the appropriation of their fields, the relocation of their homes, and the privatization of their commons—as a result of the LHWP is premised on the privileging of market-based evaluations of resources and livelihoods and the anti-democratic assumption that political and financial elites have the right to make top-down decisions that literally sell the river, and the communities that rely on it, on a calculation of increasing GDP as a measure of a nationalistic "greater good."

While there is an environmental injustice in the paradigms that support dam building in the Third World, there is also classism and discrimination within Basotho society regarding who can absorb the costs of this nationalistic project. The remote, rural highlanders' lives and livelihoods are less valued than the urban, educated elite, mostly men who hold the majority of decision-making power within Lesotho and control the limited local credibility within the global international development industry. In considering neoliberalism, nationalism, and development through the lens of environmental justice and feminist political ecology, a critique of global and national development agendas that transform communities and landscapes irreparably, while unevenly distributing the social and ecological costs and benefits of these national mega-projects, emerges. Whether large dam building is a case of elite self-deception about the real, lived consequences of mega-projects or a demonstration of the power of ideology and social distance in creating these kinds of nationalistic solutions to poverty and development, the environmental justice framework asks whether these elite decision-makers would prescribe and support mega-dams if they lived in the communities directly absorbing the costs? If the burdens and benefits of these projects were more evenly distributed and if those absorbing the costs of this project were involved in the decision-making, how might the moral and policy calculations be different? The environmental justice framework demonstrates the need to take egalitarian and democratic commitments more seriously and interrogate claims of "the common good" by carefully including those most likely impacted by a proposed policy in a conversation about their "good." An intersectionally oriented feminist political ecology pushes us to consider these questions of justice through attention to power across scales and in their interconnections. In this way both feminist political ecology and the critical analytic of environmental justice may align with its normative appeal to create more just development policies informed from the perspective of lived experience rather than a singular paradigm of economic growth and market gain.

Notes

1 Historically this has been referred to as "oscillating male migration" (Murray 1981). It refers to the cyclical movement of migrant Basotho male miners to South African mines for eight to 11 months of the year and their return to Lesotho for the remaining months of the year.

2 Women's rights advocates have long pointed to how women's legalized secondary status contributed to a culture of violence, undermined the rights of women and girls in multiple ways, and had significant impacts on their education, careers, income, health, nutrition, sexual and reproductive rights, and safety from violence, and served to legitimate other customary relations of gender discrimination (Letuka et al. 1997).

3 The unemployment rate refers to the percentage of the adult labor force (15 and older) without employment, including those seeking employment, those available to work, and those who have looked for work recently.

4 An evaluation of the LHWP by the World Bank (2010, 2) reports that:

> the distribution of Lesotho's income is one of the most unequal in Africa. Some 60 percent of the population are poor, and a third are in extreme poverty. The poverty incidence is 72 percent in rural areas compared with 27 percent in Maseru, the capital.

5 South Africa was anticipated to pay approximately US$55 million in royalties to Lesotho each year. In 2002, reports indicated that Lesotho had received about US$18 million in average annual revenues due to less than estimated amounts of water

(Hassan 2002). As of 2006, the World Bank reported cumulative royalties were 1.9 billion Maloti or US$240 million, which is slightly above projections upon initial appraisal (2010, 10). The World Bank also note these royalties have not been used toward the stated goal of poverty alleviation (2010, 9).

6 "Basotho" refers to people of Sotho origin (singular, Mosotho) in Sesotho, the primary language of Lesotho. Where necessary I have used identifiers for an English audience that would be redundant in Sesotho (e.g. Basotho people).

7 I also conducted surveys during my fieldwork in 2000–2002. These data do not speak to the analyses raised in this chapter and will be presented in future research elsewhere.

8 It is important to note that villages vary greatly in their size.

9 Due to my limited language training, I always conducted interviews in Sesotho with a translator present.

10 McMichael (2008, xxxviii) states the conventional definition of development is "nationally organized economic growth," but he argues that in the era of globalism a better definition of development is "globally organized economic growth."

References

Banerjee, D. and M.M. Bell. 2008. Environmental justice. In *Encyclopedia of race, ethnicity, and society*, ed. R.T. Schafer, pp. 443–446. Thousand Oaks, CA: Sage.

Bennett, M. 2006. Lesotho's export textiles & garment industry. In *The future of the textile and clothing industry in Sub-Saharan Africa*, eds H. Jauch and R. Traub-Merz, pp. 165–177. Bonn: Friedrich-Ebert-Stiftung.

Berg, B. 2009. *Qualitative research methods for the social sciences*. Boston, MA: Allyn & Bacon, Inc.

Braun, Y.A. 2005a. Resettlement and risk: Women's community work in Lesotho. *Advances in Gender Research* 9: 29–60.

——. 2005b. Selling the river: Gendered experiences of resource extraction and development in Lesotho. *Research in Rural Sociology and Development* 10: 373–396.

——. 2011a. Left high and dry: An intersectional analysis of gender, dams, and development in Lesotho. *International Feminist Journal of Politics* 13(2): 141–162.

——. 2011b. The reproduction of inequality: Race, class, gender and the social organization of work at the site of large-scale development projects. *Social Problems* 58(2): 281–303.

—— and M.C. Dreiling. 2010. From developmentalism to the HIV/AIDS crisis: The amplification of women's rights in Lesotho. *International Feminist Journal of Politics* 12(3): 464–483.

—— and L.A. McLees. 2012. Space, ownership and inequality: Economic development and tourism in the highlands of Lesotho. *Cambridge Journal of Regions, Economy and Society* 5(3): 435–449.

Bullard, R.D. 1990. *Dumping in Dixie: Race, class, and environmental quality*. Boulder, CO: Westview Press.

Central Intelligence Agency (CIA). 2010. *CIA World Factbook: Lesotho*. Washington, DC: Central Intelligence Agency.

Crenshaw, K. 1991. Mapping the margins: Intersectionality, identity politics, and violence against women of color. *Stanford Law Review* 43(6): 1241–1299.

Epprecht, M. 2000. *"This matter of women is getting very bad": Gender, development and politics in colonial Lesotho*. Pietermaritzburg: University of Natal Press.

Ferguson, J. 1990. *The anti-politics machine: Development, depoliticization, and bureaucratic power in Lesotho*. Minneapolis, MN: University of Minnesota Press.

Global Policy Network. 2006. Country report – Lesotho. www.gpn.org/. Accessed October 9, 2013.

Hassan, F.M.A. 2002. Lesotho: Development in a challenging environment: A joint World Bank-African Development Bank evaluation. Washington, DC: African Development Bank and World Bank.

Hussain, M. 2008. *Interrogating development: State, displacement, and popular resistance in North East India*. New Delhi: Sage.

Khagram, S. 2004. *Dams and development: Transnational struggles for water and power*. Ithaca, NY: Cornell University Press.

Lesotho Highlands Development Authority (LHDA). 1986. LHDA Order of 1986. Lesotho Government Gazette Extraordinary. Maseru, Lesotho.

Letuka, P., K. Matashane, and B.L. Morolong. 1997. *Beyond inequalities: Women in Lesotho*. Maseru, Lesotho: Southern African Research and Documentation Centre.

McCall, L. 2005. The complexity of intersectionality. *Signs: Journal of Women in Culture and Society* 30(3): 1771–1800.

McMichael, P. 2008. *Development and social change: A global perspective*. Thousand Oaks, CA: Pine Forge Press.

Murray, C. 1981. *Families divided: The impact of migrant labour in Lesotho*. New York: Cambridge University Press.

Panel of Environmental Experts (POE). 1991. Report to LHDA and the World Bank. Maseru, Lesotho: Lesotho Highlands Development Authority.

Peet, R. and M. Watts, eds. 2004. *Liberation ecologies*. London: Routledge.

Pottinger, L. 1996. Police kill striking dam workers in Lesotho. *World Rivers Review* 11(4): 3–6.

Radel, C. 2009. Nature/environment: Natures, gendered. In *International encyclopedia of human geography*, eds R. Kitchin and N. Thrift, pp. 331–336. New York: Elsevier.

Rocheleau, D.E., B. Thomas-Slayter, and E. Wangari, eds. 1996. *Feminist political ecology: Global issues and local experiences*. London and New York: Routledge.

Roy, A. 2000. *The algebra of infinite justice*. New Delhi: Penguin Books.

Scudder, T. 2006. *The future of large dams: Dealing with social, environmental, institutional and political costs*. London: Earthscan.

Sechaba Consultants. 1994. Poverty in Lesotho 1994: A mapping exercise. Government of Lesotho Working Document. Maseru, Lesotho.

Southern African hearings for communities affected by large dams. 1999. Voices of Affected Communities. www.africanwater.org/com_aff_by_large_dams.htm. Accessed April 1, 2014.

Truelove, Y. 2011. (Re-)conceptualizing water inequality in Dehli, India through a feminist political ecology framework. *Geoforum* 42: 143–152.

Tshabalala, M. and S.D. Turner. 1989. 1988 Socio-economic census of the Lesotho Highlands Water Project phase 1A areas, Volume I, Main Report. Maseru, Lesotho: Lesotho Highlands Development Authority.

United Nations Lesotho. 2007. *United Nations development assistance framework 2008–2012*. Maseru, Lesotho: United Nations.

Walker, G. 2012. *Environmental justice: Concepts, evidence and politics*. London: Routledge.

World Bank. 2010. *Project performance assessment report: Lesotho Highlands Water Project, phase 1B (Loan 4339-LSO) Community Development Support Project (Credit 3308-LSO)*. Washington, DC: World Bank.

World Commission on Dams (WCD). 2000. Dams and development: A new framework for decision-making: The report of the World Commission on Dams. London: Earthscan.

Yanow, D. 1995. Built space as story: The policy stories that buildings tell. *Policy Studies Journal* 23(30): 407–422.

3 The silent (and gendered) violence

Understanding water access in mining areas

Kuntala Lahiri-Dutt

Gender is the key explaining factor for why the social and environmental impacts of mining do not affect communities in a homogeneous manner. Deliberate neglect of gendered power dynamics within the community often engineers a shift to, or enhances, masculinities associated with mining projects in the Global South. Drawing on a feminist political ecology perspective, this chapter argues that a rights-based approach to the study of mining impacts on communities must also integrate gender considerations when thinking about changes in water quality and quantity caused by large mineral extractive operations. This argument is illustrated by coal mining-induced water conflicts in eastern India, in the state of Jharkhand,[1] particularly in the North Karanpura valley where mining has rapidly advanced in the last two decades. Consequently, access to local water resources—historically part of the local environmental commons—has declined steadily. The loss of natural water resources is felt most acutely by indigenous women who have used these water sources for generations for daily subsistence and places of social interaction. This chapter shows that decreased water availability due to large-scale mining has impacted gender roles within the family and community. However, women have risen up against collieries that undermine their existence. These dynamics are illustrated through the story told by an indigenous village woman named Shanti who describes the impacts that the entry of a mining company had on her community particularly with respect to their access to water.

To address these challenges, linkages between mining and water, then rights to land and water are discussed related to mining communities globally. The context of coal mining in India (India is the world's third largest coal producer) and in Jharkhand, a state in eastern India, is explored. Finally, the story of villagers' experiences with a coal mining company in Jharkhand is told through the eyes of one indigenous woman.

Mining wealth, mining water

Whether or not large mining projects act as engines of human and economic development in resource-rich, economically poor areas is a still-unresolved debate. Mining projects extract mineral resources accumulated under the land surface and use up some of the natural capital critical for sustaining the well-being of local communities who live in areas immediately surrounding the mining operation. In other words, large mining projects located in resource-rich areas create wealth and extract value from environmental resources for a wider market, while weakening the ecosystem that provides the subsistence base for long-established communities. Particularly in India and throughout the Global South, affected communities include marginal ethnic groups who bear disproportionate costs associated with environmental degradation. Inequities are ingrained in the very nature of the mining industry: highly capitalized, large-scale mining is connected to powerful national and global economic networks. At the same time, large-scale mining causes environmental degradation (such as deforestation, water and air pollution) and social impacts (such as cultural and occupational displacement) at the local level, where communities live and depend directly on environmental resources. Mineral resource extraction tends to 'reproduce' rather than 'reduce' inequalities in wealth and livelihood opportunities (Bridge 2004), and if women[2] and men have unequal access to and control over resources, then mining enhances these inequities.

In India, the inequity is reinforced by the state, which has complete ownership over mineral resources, and which enacts laws to ensure this primacy of rights, including rents derived from minerals in the form of royalties and lease payments from mining companies. In contrast, local communities' rights over the same resources or ecosystems are often not legitimized by laws. More importantly, in their eagerness to extract minerals, mining corporations tend to focus on individual and isolated parts of the environment while neglecting others. This view is in contrast to many indigenous communities in India that consider the environment holistically, refusing to reduce the environment into its various parts (or 'cultural flows' within waterscapes, Jackson 2005). Land, its resources (such as forests above and minerals below the surface), and water are intimately enmeshed in this life-world.

Along with the in- and out-flows of capital and the visible and invisible transfers of wealth from the local area, the extraction of minerals from the earth's surface also leads to a transfer and transformation of water resources and additional demands on local water reserves. Groundwater tables and natural water sources such as springs, rivers, and lakes are depleted by mining activities. Mining pits often extend well below the groundwater table, and as the mines need 'dewatering' (elimination of water by pumping), the overall groundwater table of the area may fall, resulting in changes in rivers' flow, a drying up of natural springs, or drying up of domestic water wells. Mining requires significant quantities of water for processing and transporting ores and wastes. Even washing heavy equipment and spraying water to suppress dust requires large quantities of water. Additionally, certain processes, such as blasting rocks or material removal, lead to water diversions from ecosystems (DRET 2008). Mining settlements also demand water supplies for human use. Due to these large water input needs, mining companies located in wet or dry climates generally store water in large dams or reservoirs to ensure a steady supply of water.

Mining processes also often release water back into the ecosystem as wastewater contaminated with various chemicals and crushed rocks. Some of these chemicals are foreign to the local rocks or were previously bound up in the rocks. It is not uncommon for mining companies to further disrupt local ecosystems by releasing wastewater into local rivers, lakes, and oceans. The 'tailings' in mines contain waste slurries comprising minerals and other wastes like chemicals that are dangerous to human health. Tailings can cover extensive areas and if not contained, can contaminate surface water sources and the water table (Bebbington and Williams 2008).

Water is important for the mining process and has many ecological and health implications, but it is also a crucial element for the subsistence and cultural identity of communities near mining areas. Furthermore, almost all water-related chores—from the identification of clean and reliable sources to daily water collection and storage—are allocated to women as gendered tasks (Mishra 2014). Therefore, women in many local and indigenous communities have materially based and intimate relationships with water, not just an ambivalent closeness to water by virtue of their being women and for their reproductive roles. The mining industry has focused on water quality, ignoring these more complex issues of gendered water access so closely related to human rights.

A gender perspective is critical to truly understanding the impact of large-scale mining, because women and men have different social roles, rights, and opportunities, and are differentially affected by any major change in land use, particularly when land and water resource ownership is temporarily transferred to a non-local investor or corporation.[3] Gender is also a major variable in water rights as they are connected to land ownership and land use rights. Throughout much of Asia, land is most often owned by men and is usually passed on to sons. Most marriages are virilocal, particularly in India, even among indigenous communities. Since young women go away to their husband's village, the 1956 equal inheritance act does not 'work' (Agarwal 1994; Rao 2008).

Gender is also pertinent to this debate for reasons other than the materialist argument.[4] Large mining projects in remote and isolated areas introduce new gender cultures—primarily market orientation and cultures of hegemonic masculinity—that emerge in the inorganically imposed mining settlements located at the resource peripheries. The new industrial landscape pervades not just the physical aspects of the environment, but also the social and cultural landscape, producing a new, commercialized, and reductionist view of the natural environment and its individual parts, such as land, water, and trees. These elements of the environment are valued individually in discussions on financial compensation, which in most cases take place between men from the mining corporation and men from the community. The new commercialized landscape attributes a domesticity to women by socially constructing and locating women within the home. There are many examples of this both within India and in other world regions. For example, in Bolivian mining communities, Nash (1979) showed how women were squarely placed within the context of the home as a male miner's wife. Such a 'housewifization' of all women by the new political economy, for all practical purposes, entrenches women in their reproductive roles, as *only* domestic subjects.

The compartmentalization of women and men into reproductive and productive boxes also alters traditional gender roles in which women might have exercised significant authority over production and household food security, undermining their traditional status and prestige. In most indigenous communities, women and men have different positions and status; they inhabit different spaces, perform different functions, and play different roles. Not all differences are hierarchical in nature; often, significant amounts of social prestige and status are associated with roles played by women. New gender cultures, however, replace conventional symbols of power and prestige with cash, which is generally accessed only by men, making women bear gender-selective mining impacts. Thus, mineral extraction leads to a degraded ecosystem and new economy, which drive women to the margins.

Water: what right is right?

This chapter acknowledges the principle of 'primacy of rights' (as elaborated by Gleick 1998) with relation to water and gendered access to water resources. The language of rights has ascended to the international water policy domain and not unsurprisingly, has gradually been entering the highly capitalized and corporatized extractive industries sector. A 'water right' gives its holder authorization to extract water from a particular source and includes social privileges and obligations (Zwarteveen and Meinzen-Dick 2001). The legitimacy of right holders' claims to water is linked to social relations of authority and power. Mining corporations are more economically and politically powerful than local communities and state law recognizes their rights to water and resources, whereas the claims of indigenous communities are often customary. This is particularly true for women who do not own the land, but have conventionally used the water.

Communal user rights over water in the Global South are gradually giving way to individualistic use rights (Achterhuis et al. 2010). The question of whether the right to water is an economic or social right still assumes significance when more than one type of use or users compete over limited supplies, or where water resource availability declines due to pollution and groundwater overdraft. Whenever competition for water becomes intense, the less powerful generally lose out.

In mining areas where industry imposes new demands on limited water supplies, Kemp et al. (2010) identify four types of water rights: legal, cultural, universal, and distinct human rights. As noted by Bruns and Meinzen-Dick (2000) there are significant differences between various kinds of rights, such as user rights and ownership rights. Kemp et al.'s argument, therefore, can be extended to include the question of justice. The establishment of private rights to water in mining areas raises ethical questions regarding the equitable allocation of a resource that is, under the cultural and religious traditions of many populations, a common heritage. In mining areas, rights to water devolve to the company by virtue of its land lease. The issue assumes practical/operational dimensions when mine owners assume control of water sources within their lease boundaries and drill wells to redistribute 'their' water. They take agricultural land out of production through such mal-distribution, destroy natural water sources, and pollute waters. In this situation, do the disenfranchised and displaced populations have any 'right' to protest against the reallocation of water or claim a share in water sales? What rights do women retain when water rights are reallocated? Furthermore, if public or individual 'non-right holder' rights remain, how should they be protected and given voice? The latter is particularly relevant in thinking of gendered water rights, because often rights used by women fall into the category of unwritten, customary, and legally unacknowledged user rights.

The complexities surrounding mining, water, and gender, therefore, can be linked to environmental, social, and cultural justice. The consideration of water as a human right arose primarily to counter right-wing efforts to treat water as purely an economic good (Iyer 2008), clearly bringing into focus the contradictions between economic efficiency and basic welfare, or human values and water rights. 'Human rights' refers to rights and freedoms recognized by the global community, protected by international legal instruments, and which are universal—the birthright of every woman, man, and child (Lauren 2011). States are bound by international obligations to realize human rights for their citizens. Recognition of access to water as a human right within international policy commitments in the mining sector has yet to come about. Generally speaking, the 'right to water' is still widely debated and is closely associated with rights to life and health (Sultana and Loftus 2012; Whiteford and Whiteford 2005).

A human rights framework related to justice would ensure that social (including gender) and power inequities are not enhanced by environmental degradation, and that physical and occupational displacements are addressed equitably (Schlosberg 2007). Mohai et al. (2009) place these concerns within the framework of environmental justice, which spreads far beyond the lack of political participation, to encompass the failure to recognize customary water rights at the individual and/or community levels. Since water is crucial for human well-being, mining companies that fall within the ambit of human rights laws need to ensure women's access to water.

Mining and water share the same turf, the surface land, where the battle of rights is fought. Thus, land and water appear to be different aspects of the same natural capital necessary for the well-being of indigenous communities (Lansing et al. 1998; Strang 2009). Land rights legislation in most countries has transitioned from feudal notions of rights to individual rights over cultivated land; with water rights there has been a shift away from community ownership and use, towards state ownership and individual use (Ivison et al. 2000). Water, and access to it, relates to the livelihoods of poor people—indigenous communities usually form a large section of these poor (Colchester 1995a, 1995b; Verma 2001). When women comprise the most burdened group within the mining communities, with the least voice and rights, the interface of mining, water, and gender rights poses a serious problematic for feminist political ecologists.

Although the contestations between mining and water are well-recorded, primarily from the physical environmental point of view (Chatterjee et al. 2010; Luís et al. 2009; Singh et al. 2008), existing literature is relatively silent on gender issues. One exception is a study of women who organized against water pollution from copper mines in Cananea, Sonora, Mexico (Browning-Aiken 2012; Browning-Aiken et al. 2003). From the more technical questions of bringing water to the mines, to mining water use and impacts of mining on water (such as pollution and lowering of groundwater tables), to mine tailings and to reductions in fish populations (Chaulya 2004), even the staunchest mining engineer recognizes water issues as closely linked to their business. Mining regions often suffer from increasing impoverishment of indigenous communities (see Ali 2004 for a comparative study of environmental negotiations between mining companies and native populations in Canada and the United States). Strang (2009) showed the conflicts between indigenous values of water and those held by Australian mining companies. Nakashima (2001) noted how coal extraction in Navajo and Hopi territories in the southwestern United States resulted in the drying up of wells and springs used for long periods by the community. More recently, Imai et al. (2009) described the resistance by an indigenous group in Guatemala to a Canadian-owned mining company for violating their customary water rights on the basis of Guatemalan law.

In the intense debates over water resources in India, indigenous communities have indeed played a major role since the early days (Ganguly-Thukral 1992). Padel and Das (2010) have equated the dispossession of indigenous communities by large mining corporations to 'cultural genocide.' Upadhyaya (2003) has brought up the question of customary rights over small and local water sources such as tanks (ponds). He pointed out how the highest legal authority in India, the Supreme Court, in a recent judgment, noted that customs are only a *source* of law and can only become such a source when they are recorded in statutes or are recognized by courts. In India, formal laws do not yet recognize the customary rights of indigenous communities. Arguments on and for customary rights of indigenous communities over local land and water will have to be placed within the gambit of formal laws to be taken seriously. Upadhyaya (2002) also showed that a contemporary court case brought attention to traditional rights of a fishing community to catch and carry fish. No such attention has yet been paid to the issue of compensation to communities for lost natural water sources due to mining in general and coal mining in particular. However, one non-governmental report in 2008 (the *Sixth Citizen's Report*, published by the Delhi-based Centre of Science and Environment) did recognize the water pollution in areas where land was traditionally held by indigenous communities, demonstrating the need for the government to take this matter more seriously (Bhushan and Hazra 2008).

In the next section, I briefly discuss coal mining in India before dealing with the water traditions of Jharkhand and coal mining's impact.

Water meets land and its peoples on coal tracts

In 1991–1992, the government of India adopted radical and structural reforms, and embraced neoliberal economic policies. This marked a watershed in post-colonial India's economic history, which had been characterized by a loose model of Soviet-style central planning, with time- and sector-specific targets and goals. The rapid expansion of coal mining to produce thermal power for the industrial and urban sectors was a key plank within the suite of programs adopted by the state (Dias 2005; Fernandes 2012). This led to a sudden rise in the number of coalmines, particularly in the resource-rich, but economically deprived state of Jharkhand in eastern India. The new coalmines in Jharkhand, mostly open-cut, coking coalmines used in steel production, have a very large environmental footprint leading to the rise of vocal anti-mining civil society groups resisting the rapid degradation of the environment (Energy Information Administration of the U.S. 2014; Vagholikar et al. 2003). Over 60 percent of the greenhouse gas emissions (mainly carbon dioxide, nitrogen oxides, and chlorofluorocarbons or CFCs) of India are due to the combustion of coal for power plants and industrial processes. Large energy losses are involved in transmission and distribution of electricity generated from coal-based power plants in India. These processes and greenhouse gases are major contributors to global warming (India Times 2014; Raghuvanshi et al. 2006).

The expansion of coal mining has given rise to three main overlapping questions related to resource ownership. First, the mineral resources often occur on indigenous-owned land or on village commons, where customary property rights to water and land have been held. This leads to questions over the formal acceptance of customary rights. Second, the overall legal structure used by the Indian government to define resource rights is of colonial vintage, and thus raises questions of the laws' adequacy to meet contemporary needs. Third, the rampant corruption and informal nature of economic transactions cast a shadow over the mine planning, compensation, resettlement, and rehabilitation processes. The use of force in relocating poor people in colliery regions is not uncommon. Even current environmental laws can become irrelevant in the face of these corrupt and violent factors.

The Indian Ministry of Coal does not maintain reliable, comprehensive data on the number of people impacted or displaced due to coal extraction. Mining displaced communities (DPs, also known as project-affected people or PAPs) are treated as a homogeneous group, without attention to diversity within these communities. Furthermore, many new open-cut coal mining projects exclude indigenous people, especially women, from the social and economic benefits they produce, whether due to their perceived lack of skills, formal education, or ownership and control over land and water. There are no gender-segregated data from mining companies under the Environmental Management Plan. The social and gendered consequences of specific mining projects may depend on the local context, but the most common outcome is impoverishment and pauperization through displacement.

A gender analysis framework is necessary to reduce impoverishment risks for displaced communities. The social and economic impacts of mining are gendered in the sense that women are less likely to share in the benefits such as wage employment, and more likely to bear the burden of risks such as loss of sustainable livelihoods. Moreover, the loss of environmental resources and changes in social fabric are closely interrelated. Indian women, especially those in villages, do not have legal rights over land, and are rarely titleholders (Rao 2008). Compensation processes for land reclamation from the mining companies usually only accept an adult man as head of household, failing to consider women's needs and requirements. Hourly or wage jobs, if any, are usually awarded to men, and women are forced either to stay at home as housewives, or to look for wage labor positions far from their communities (Fernandes 1998; Ganguly-Thukral 1996). In circumstances where men are seen as more legitimate wage earners, women left in the community to care for children while their husbands are away can find themselves overwhelmed with new responsibilities (Fouillard 2003).

Jointly with colleagues in academia and civil society organizations, I have been researching the gender-specific impacts of coal mining since 1994–95. An early study in Jharkhand showed that women were more affected by displacement than men because physical amenities and services were rarely provided at relocation sites, and women were more dependent on them than men due to their many gender-specific domestic responsibilities (Lahiri-Dutt 2006). In 2006 and 2011, we conducted an intensive survey of the challenges facing women in resettlement colonies (Ahmad and Lahiri-Dutt 2007; Lahiri-Dutt and Ahmad 2014). We also participated in the Oral Testimony Project in 2002, which recorded oral testimonies of indigenous communities displaced by coal mining operations (Bennett and McDowell 2012). We visited with these individuals again in 2012 to track the changes in their livelihoods and well-being over the last decade. The following sections draw on these research processes. The map in Figure 3.1 indicates the location of the area discussed in this chapter.

Water traditions in Jharkhand and the impacts of mining on them

There are several sources of water that indigenous communities commonly use: rivers, wells, ponds (tanks in Indian English, or *talaos*), and natural springs (*darhis* or *jharnas*). For the *adivasis* (aboriginal peoples) of the land, water has always been a natural gift, a part of the landscape and source of life. It has always simply been there—in rivers, in streams, in springs, and in wells. Nature in this part of India has often been whimsical and linked to unreliable monsoon weather patterns that often lead to diminishing springs and drying up of rivers in the summer. However, sooner or later the springs brought back the life-giving water and rivers flooded. These changes were never quite certain, and sometimes caused hardships and forced groups to make adjustments to a reduced flow, but were taken as part of the life process.

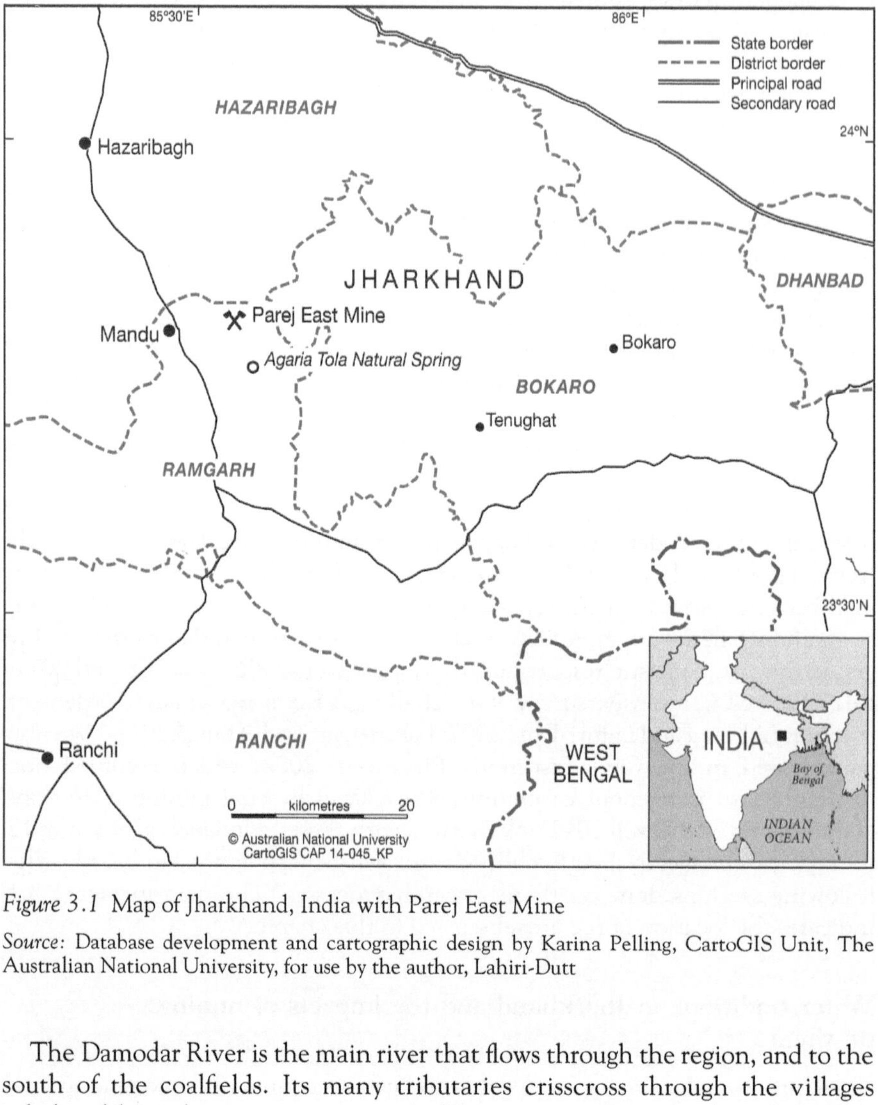

Figure 3.1 Map of Jharkhand, India with Parej East Mine

Source: Database development and cartographic design by Karina Pelling, CartoGIS Unit, The Australian National University, for use by the author, Lahiri-Dutt

The Damodar River is the main river that flows through the region, and to the south of the coalfields. Its many tributaries crisscross through the villages inhabited by indigenous communities. These seasonal streams have been the main sources of water for bathing, washing animals, washing clothes, irrigation of fields, and fishing. Women dig holes in the sandy riverbed to collect filtered water for the household in summer. The vital importance of the streams to indigenous communities cannot be underestimated.

Natural springs (*darhis*), once the main sources of water for indigenous communities, have now become scarce due to encroachment. Often a fence is built around them for protection, and women keep the surroundings clean. A large spring is called a *jharna*. The story of one such *jharna*, that of Agaria Tola, which is a hamlet near Pakoddih of Burugutu, as told by Shanti Murmu, supports

Figure 3.2 Women and girls collecting water, Jharkhand, India
Photo credit: Kuntala Lahiri-Dutt

the arguments made by this chapter. Wells can measure from 2.5m to 4.5m in diameter, with depths of 4.5m to 12m, lined with bricks or stones. They are built either from villagers' own resources, or with funding from government rural development projects (see Figure 3.2).

A *talao* is made by damming a depression or rivulet with an earthen wall, or simply by excavating low-lying ground. Local communities have traditionally cared for the *talaos*, and the water collected in *talaos* is for community use. Many *talaos* dry up in the long summer months, leaving only the *darhis* and *jharnas* to provide drinking water. Hand-pumps are provided by the government and are favored by villagers, particularly women, because this saves them from lifting heavy buckets over long distances during the dry months. However, the water from these pumps is not reliable; sometimes the water is clean, but sometimes it is polluted by minerals. Moreover, the pumps tend to break down after a year or so, and become useless, since resources for repairing them are not readily available.

It is well known that open-cut coal mining impacts water resources, directly affecting water quantity and quality and indirectly affecting the water table through increased demand from the influx of people from other areas who move to the region for mining jobs. However, another notable change in water relates to how water is perceived by mining companies, indigenous peoples, and migrants to mining communities. Unlike the indigenous peoples, the *dikus*, or outsiders, do not see water as a sacred resource to be celebrated through symbolic rituals. A mining company treats water sources within underground mines as part of the 'overburden' to be 'removed.' Water is now brought in by tanker trucks from outside the area, thus making communities dependent on external water suppliers. Whereas individual well or pond owners receive compensation from the company, the destruction of natural water resources (including rivulets and springs) on common lands is not even noted as damage for which compensation needs to be given to local communities.

Crisis of drinking water

Around coal mining villages, acute drinking water crises are evident, generally throughout the year, but more intensely during the dry months from December to June. This crisis is primarily due to groundwater depletion since the advent of mining. However, it is not apparent in groundwater data collected seasonally by the State Groundwater Board, because their established wells are generally located away from the coalmines and new wells have yet to be drilled. Fieldwork revealed that open-cut mining has also led to surface and ground water pollution (see Figure 3.3). Lower-caste families bear a greater burden of water shortage. Before the advent of mining, for most women, safe drinking water was available in the villages within a one-kilometer radius. Now, women are forced to walk longer distances, trekking across the mine to other villages. Lack of water has a direct bearing on reduced food security of poorer families who previously depended heavily on the water and agricultural land now occupied by mining companies. The quality of water has also changed; Malti, resident of a village named Benti, observed:

> For washing and bathing, we use water from the mines. This water makes our skin itch. Our body becomes sticky after using the mine water. When there were no mines here, we used water from the Dahenchawa River for drinking as well as for washing and cleaning. Now, even the little water that is left in that river has become full of coal so we cannot use it.

Those who can afford it buy water from a local entrepreneur who extracts water from the nearby Damodar River and pipes it, untreated, into their houses. Other, better-off families have also installed their own tube wells (borewells) or deepened existing private wells. Such differential access to safe and clean water sources has led to internal strife within the community and has increased social inequalities.

Company efforts: minimal

There have been minimal attempts to remedy the situation and to provide substitutes for lost water sources. Pollution monitoring/control is done at a

Figure 3.3 Open coalmines in Jharkhand, India
Photo credit: Nitish Priyadarshi

number of water monitoring points where the coal company collects water samples and sends them to the State Pollution Control Board. However, local experts and civil society organizations have expressed doubts about the accuracy of the sampling, as the results are not publicly available (Priyadarshi 2003). The mining company also provides replacement wells, but they also go dry in summer. Local communities have attempted several substitutes. The mining companies sometimes drill deep borewells for their employee colonies, but these have been a bone of contention because the indigenous Santhal community claims this land as theirs. Eventually in 2007, the indigenous community members arrived with bows and arrows, broke the structure, and filled the borewell with sand and rocks. Currently, efforts are being made to control run-off from washeries and workshop areas by garland drains, but water shortages lead villagers to use the contaminated water for bathing and washing and women carry it home for household use. The mining company provides some water in tankers during extreme water scarce periods in summer; this water is usually pumped out from abandoned quarries and is not of very good quality, but is nevertheless used for washing and cooking.

Let us understand this experience through Shanti's eyes. She is a Santhal and was a resident of Agaria Tola village, which once had a spring of the same name, but both no longer exist. Shanti is a widow of about 45, and lived in this village for about 30 years. Shanti's income source is a daily agricultural wage from local fields. Shanti's narration was recorded during the first phase of the Oral Testimony

Project in the area in 2002. She focuses on a natural spring, and describes how women tried to protect it, and how the mining company used force to destroy it. I interacted with Shanti through Father Tony Herbert of Prerana Resource Centre and Bina Stanis of Chotanagpur Adibasi Seba Samiti (committee). She spoke in Santhali language which was translated into Hindi and English.

Shanti's story of Agaria Tola

We women from our Agaria Tola and neighboring hamlets would all come to this spring. We were maybe about 100 families in all, not as many as in the colonies, not rich like them too. But the spring was precious to us, so we affectionately called it *naihar*.[5] We met there in the spring, morning and evening. It was a place where we gathered every day to bathe our children and ourselves. There were several snakes there in the spring, and we treated them as our own. For us, the spring was not just a place to collect fresh water, but was a place to meet.

One morning, the TISCO[6] Company bulldozed that spring, and with it the large *asan* tree[7] next to it . . . I can still see the spring and the *asan* tree next to it, if I close my eyes.

I ask, did the *sarkari* (government) officers know about the spring and what it meant to us, local villagers? Did they include its value in the cost of their mining? If they counted its value, it would be many thousand times more valuable than all the coal they will mine from there. That spring has been with us, like our mother's place, for many, many years. The company saw the coal only, not the need of ordinary people like us for water of the spring. Tell me, can we go and cut off the company's water? If we destroy the water supply of the company colony, there would be criminal charges against us. So if the company destroys our *naihar*, how is that something different?

As the spring was on the edge of the mine lease, our leaders proposed that some area be marked off to preserve the spring. They also proposed that some area be left as its suitable catchment area. The officers did not pay heed to our demands. So we decided to put up resistance to the company . . . All of us villagers went into the mine several times and even stood in front of the machines to stop their work. Each time we protested, the company called in security officers and police . . .

Even our ancestors are unhappy with the things happening in our land. Their curse broke the machine once. The story is like this: TISCO had brought in a large new shovel machine. Two days later, it broke down at the site where the spring was. Somar, our village leader, said that it was the deity of their spring whose curse caused it to stop working . . . The super power machine sat there for a month. Next month, one of the TISCO officers, who had repeatedly come to the village and told the villagers to let go of the spring, met with a bad road accident and he and his family were killed. We believed that this too was the curse of our deities whose abode in the spring was destroyed for mining. Even TISCO began to believe that there was some

> truth in what we said, so they offered to give us a big feast to appease our gods, so that they could continue to extract coal.
>
> The company then dug a well for us, but it became dry the next summer. The company also provided water through tanker trucks. We villagers asked, how can an unknown driver, bringing and pouring water into cement troughs, be a substitute for a living spring? Even then, it was an old oil tanker, and the water smelt of kerosene.

Shanti's story takes an interesting turn at this juncture. The Central Coal Fields Limited (CCL), in order to avoid spending money on compensation and rehabilitation, decided not to mine under the houses, but instead cut around the edge of the village, leaving the homes. In the process, the fields and the forests were completely destroyed, and the water table was lowered throughout the area, leaving no way for women to fetch water or for men to carry on with farming. Due to its location in the buffer zone of the mine, Agaria Tola came under the World Bank's Coal Sector Environmental and Social Mitigation Project. The villagers were given 'development assistance,' which took the form of renovation of a pond, six months of adult education classes, and alcoholism and development-related consultations. Women resisted. In Shanti's words:

> We appealed to CCL to give us drinking water, but they paid little attention to us. We took protest groups twice to their offices but the officers did not relent. Consequently, we women went into the mine and stood in front of the machines, and stopped the mining work. It was only for a day that we were able to stop the mining. Still, we felt that it was a big achievement for us, to be able to stop the machines that killed our spring.

I will end Shanti's story with a few words she told us in 2012 in a follow-up interview as part of our Oral Testimony Project. At that time, Shanti and her neighbors had been hanging precariously onto their homes to resist relocation to a colony. There were no more forests or cultivable land around, and the water table had fallen far below the ability of hand pumps to draw out water. Most people depended on colliery tanker trucks supplying inadequate and infrequent quantities of poor-quality water. For their livelihood, most men, women, and children had turned to 'coal collection,' scavenging on open-cut mines and abandoned pits, and digging up any available land for coal that they carried with bicycles to customers located in nearby towns. Shanti regretted the ways of life that will never come back, and said:

> I ask, are we also not citizens of this free country? Are we too not to live in human dignity? Must we live like animals in our own village? Why is it that the only language the mining company understands is that of protest? Why does the mining company threaten us with police action for taking a bit of their coal? I ask, if they can destroy our water supply, how will we live? Tell me, how shall we get our *naihar* back?

What can feminist political ecology contribute to water issues in mining areas?

Shanti's story offers a glimpse of how women in long-established village communities envision water, part of their local survival base, which then gives them cause to resist a large mining corporation. It also highlights the uneven power relations between commercial actors and indigenous people, between dominating and marginalized ethnic groups, and between those who are more powerful and those who are dispossessed of their rights to land and water. However, her story also provides an actor-oriented perspective, where indigenous women (and men) resist the enormous constraints on them and protest against exploitation. She demonstrates that when entire indigenous communities are faced with the onslaught of radical and exploitative changes in their lives, it is often women who zealously guard their rights over the sources of water that are essential for their survival.

The government, on the other hand, acknowledges only ownership rights. Mining companies presume that all natural resources contained within their leasehold boundaries are part of the deal, and give little recognition to prior possession and use by communities.

Shanti's story also points to gross inequities in rapidly modernizing India in which rural women's demands for local and natural water resources in coal mining areas are given little or no importance. These unequal social relations of urban–rural, or accumulation–subsistence economies, represent a power imbalance that is reinforced by the state establishment and authorities. Truelove's examination of urban slum women's access to water in New Delhi, India utilizing a feminist political ecology framework of analysis similarly shows that men and women and people of different social classes experience the state and the law differently especially in relation to rights and resources (Truelove 2011). Added to this, in Shanti's community the private sector in the form of the mining company induces inequities related to rights and resources: the company officials receive secure and safe water supplies, but villagers receive poor-quality water in tankers. Caste inequalities of Indian society are also reflected in this power imbalance: government and mining company officials usually belong to upper castes whereas village communities belong to indigenous and *dalit* (previously termed untouchables and among the lowest caste) groups.

Conflicts between customary and statutory rights in a coal-mining region reach their best expression over water, which is the essence of life for a large number of rural women. Neither the mobilization by women, such as in the spontaneous protest narrated here by Shanti, nor the claims by women to natural water sources have any legal validity in a court of law. As the women were only spring users, and no individual could legally be recognized as its owner, no monetary compensation was paid to those who used the water. Women's claims to the spring inhabit a different, moral domain from the one inhabited by the mining company. For the women, land, water, and forests are enmeshed in society. The mining companies only value one particular aspect of the environment—the minerals that lie underground.

Stories of such protests, even those small in scale, are significant to feminist understandings of gendered resistance and environmental justice. They are at the heart of the way we think about feminist political ecology as the overarching analytical framework. It is through these stories that one begins to fully reveal gendered knowledge, gendered environmental rights and responsibilities, and gendered environmental politics and grassroots activism. These are the three critical pillars of the 'science of survival' that feminist political ecology deals with (Rocheleau et al. 1996, 2). We see how indigeneity intersects with sex to make women like Shanti key actors in everyday acts of resistance.[8] Most importantly, we begin to productively interpret how broader inequities ingrained in resource governance systems lead to gendered consequences of mining with regards to water. As geographers, we note that the gendered politics over water rights operate at the most microscopic scale, but are also intricately linked to global ecological changes in and economic conflicts over water resources.

Acknowledgments

I am indebted to Shanti for sharing her time and for telling her story, and to the communities of Agaria Tola and other villages of Hazaribagh area. I also thank Dr. Nesar Ahmad of the Institute of Budget Policy and Ms. Radhika Krishnan of Jawaharlal Nehru University for their research assistance, and Fr. Tony Herbert of Prerana Resource Centre and Bina Stanis of Chotanagpur Adibasi Seba Samiti for helping me narrate Shanti's story.

Notes

1 Jharkhand presents a special case among the Indian states. This state was carved out in 2000 from territory that belonged formerly to the state of Bihar, but was dominated by a number of indigenous communities who struggled to get their voice heard through the establishment of a separate state. In mineral-rich, forest-clad Jharkhand, land-specific laws make tribal land inalienable, that is, non-transferable to non-tribals.

2 I use the term 'women' while acknowledging that they do not necessarily comprise a homogenous group in any context whatsoever, and do not exhibit the same concerns, practices and patterns of activities and interests, assets and rights.

3 So much has been written on the gendered impacts of mining that I avoid getting into that literature within the limited space of this chapter. Suffice to say that these impacts range from erosion of the physical subsistence base to social and cultural changes such as changed notions of authority and interpersonal power equations both at home and within the communities (Macdonald and Rowland 2002). Women are affected by lack of access to assets and resources, as well as by the increased cash flows into local economies, predominantly into the hands of men. The gendered impacts often cut across class and race, but those women (and men) who often are already at a disadvantaged position within the socio-economic hierarchy are generally intensely and adversely affected by large-scale mining (see Eftimie et al. 2009). Again, the gender-selective impacts have been noted both in economically well developed countries such as Canada, as well as in less developed countries with smaller economies. (I have outlined these impacts in Lahiri-Dutt 2012.)

4 Here I must admit that I do not subscribe to the dualism in feminist approaches to the environment in the materialist-vs.-essentialist arguments put forth by Bina Agarwal (1992) and Vandana Shiva (1988, 1991). In my view, the dichotomy was unproductive theoretically and practically since much of Shiva's work can be interpreted as going beyond linking women purely to reproductive roles of rear and nurture. I have used the term to emphasize my perspective of *changing gender roles* of women and men in mining areas for the purpose of this chapter.

5 *Naihar* means 'mother's home.' This is the place a woman goes to on certain occasions, for example, to give birth to her first baby, or to seek peace when she is under tension due to family responsibilities, or if she just needs a break from her duties in her marital home.

6 TISCO stands for Tata Iron and Steel Company Limited, a major industrial corporation in India.

7 *Asan* trees are a local shade-giving species that often house the Santhal deity, the *sarna*.

8 Here I am using James Scott's 'everyday forms of resistance' through a gendered lens (see Scott 1985).

References

Achterhuis, H., R. Boelens, and M. Zwarteveen. 2010. Water property relations and modern policy regimes: Neoliberal utopia and the disempowerment of collective action. *Out of the mainstream: Water rights, politics and identity*. R. Boelens, D. Getches, and A. Guevara-Gil (eds). London and Washington, DC: Earthscan, 27–56.

Agarwal, B. 1992. The gender and environment debate: Lessons from India. *Feminist Studies* 18: 119–158.

——. 1994. *A field of one's own: Gender and land rights in South Asia*. Cambridge: Cambridge University Press.

Ahmad, N. and K. Lahiri-Dutt. 2007. Missing concerns: Engendering mining displacement in Jharkhand. *Gender, Technology, Development* 10: 313–339.

Ali, S.H. 2004. *Mining, the environment and indigenous development conflicts*. Tucson, AZ: University of Arizona Press.

Bebbington, A. and M. Williams. 2008. Water and mining conflicts in Peru. *Mountain Research and Development* 28: 190–195.

Bennett, O. and C. McDowell. 2012. *Displaced: The human cost of development and resettlement*. New York: Palgrave Macmillan.

Bhushan, C. and M.Z. Hazra. 2008. *Rich lands poor people: Is "sustainable" mining possible?* State of India's environment: A citizens' report. New Delhi: Centre for Science and Environment. www.scribd.com/doc/136875876/State-of-india-s-Environment-A-Citizens-Report-Rich-Lands-Poor-People-Is-sustainable-mining-possible. Accessed May 15, 2014.

Bridge, G. 2004. Contested terrain: Mining and the environment. *Annual Review of Environmental Resources* 29: 205–259.

Browning-Aiken, A. 2012. Disjuncture between economic policy and sustainable use of natural resources. *Neoliberalism and commodity production in Mexico*. T. Weaver, J.G. Greenberg, W.L. Alexander, and A. Browning-Aiken (eds). Boulder, CO: University Press of Colorado, 115–140.

——, R. Varady, and D. Moreno. 2003. Water-resources management in the San Pedro basin: Building binational alliances. *Journal of the Southwest* 45(4): 611–632.

Bruns, B.R. and R. Meinzen-Dick, eds. 2000. *Negotiating water rights*. New Delhi: International Food Research Policy Institute, Vistaar Publications.

Chatterjee, R., G. Tarafder, and S. Paul. 2010. Groundwater quality assessment of Dhanbad district, Jharkhand, India. *Bulletin of Engineering Geology and the Environment* 69(1): 137–141.

Chaulya, S.K. 2004. Water resources monitoring in a mining area in India. *Environmental Monitoring and Assessment* 93: 69–89.

Colchester, M. 1995a. Some dilemmas in asserting indigenous property rights. *Indigenous Affairs* 4: 5–7.

——. 1995b. Indigenous people's rights and sustainable resource use in south and southeast Asia. *Indigenous peoples of Asia*. R.H. Barnes, A. Gray, and B. Kingsbury (eds). Monograph and Occasional Paper Series, No. 48. Ann Arbor, MI: Association for Asian Studies, 50–60.

Dias, X. 2005. World Bank in Jharkhand: Accountability mechanisms and indigenous peoples. *Law, Environment and Development Journal* 1: 73–79. www.lead-journal.org/content/05071.pdf. Accessed January 15, 2014.

DRET (Department of Resource, Environment and Tourism). 2008. *Water management: Leading practice sustainable development program for the mining industry*. Canberra: Commonwealth of Australia.

Eftimie, A., K. Heller, and J. Strongman. 2009. *Gender dimensions of the extractive industries: Mining for equity, extractive industries and development*. Washington, DC: World Bank.

Energy Information Administration of the U.S. 2014. Analysis briefs-India. www.eia.gov/countries/cab.cfm?fips=in. Accessed July 7, 2014.

Fernandes, A. 2012. *How coal mining is trashing tigerland*. Greenpeace Report. Bengaluru: Greenpeace India Society. www.greenpeace.org/india/Global/india/report/How-Coal-mining-is-Trashing-Tigerland.pdf. Accessed March 1, 2014.

Fernandes, W. 1998. Development induced displacement in Eastern India. *Antiquity to modernity in tribal India*, Vol. 1, S.C. Dubey (ed.). New Delhi: Inter-India Publications, 217–301.

Fouillard, C. 2003. *Refugees and forced displacement: International security, human vulnerability, and the state*. Tokyo: United Nations University Press.

Ganguly-Thukral, E. 1992. *Big dams, displaced people: Rivers of sorrow and rivers of change*. New Delhi: Sage.

——. 1996. Development, displacement and rehabilitation locating gender. *Economic and Political Weekly*, June 15.

Gleick, P. 1998. The human right to water. *Water Policy* 1: 1–25.

Imai, S., L. Mehranwar, and J. Sander. 2009. Breaching indigenous law: Canadian mining in Guatemala. *Mining: Global experiences*. C.J. Krishna (ed.). Hyderabad: The ICFAI Press, 111–158.

India Times. 2014. Opting for coal-based energy 'wasteful approach' by government: IPCC. Economic Times. http://articles.economictimes.indiatimes.com/2014-06-28/news/50929567_1_ipcc-pachauri-intergovernmental-panel. Accessed June 30, 2014.

Iyer, R. 2008. National and regional water concerns: Setting the scene. *Water first: Issues and challenges for nations and communities in South Asia*. K. Lahiri-Dutt (ed.). New Delhi: Sage, 3–30.

Ivison, D., P. Patton, and W. Saunders. 2000. *Political theory and the rights of indigenous peoples*. Cambridge: Cambridge University Press.

Jackson, S. 2005. Indigenous values and resource management: A case study from the Northern Territory. *Australasian Journal of Environmental Management* 12: 136–146.

Kemp, D., C.J. Bond, D.M. Franks, and C. Cote. 2010. Mining, water and human rights. *Journal of Cleaner Production* 18: 1553–1562.

Lahiri-Dutt, K. 2006. Mainstreaming gender in the mines: Results from an Indonesian colliery. *Development in Practice* 16(2): 215–221.

——. 2012. Digging women: Towards a new agenda for feminist critiques of mining. *Gender, Place and Culture: A Journal of Feminist Geography* 19: 193–212.

—— and N. Ahmad. 2014. Gender in coal mining induced displacement and rehabilitation in Jharkhand. *The coal nation: Histories, politics and the ecology of coal in India*. K. Lahiri-Dutt (ed.). Aldershot: Ashgate, 221–241.

Lansing, S.J., P.S. Lansing, and J.S. Erazo. 1998. The value of a river. *Journal of Political Ecology* 5: 1–21.

Lauren, P.G. 2011. *The evolution of international human rights: Visions seen*. Philadelphia, PA: University of Pennsylvania Press.

Luís, A.T., P. Teixeira, S.F.P. Almeida, L. Ector, J.X. Matos, and E.F. da Silva. 2009. Impact of acid mine drainage (AMD) on water quality, stream sediments and periphytic diatom communities in the surrounding streams of Aljustrel mining area (Portugal). *Water, Air, and Soil Pollution* 200(1–4): 147–167.

Macdonald, I. and C. Rowland, eds. 2002. *Tunnel vision: Women, mining and communities*. Fitzroy: Oxfam Community Aid Abroad.

Mishra, P.P. 2014. Water worries in a coal mining community: Understanding the problem. *The coal nation: Histories, politics and the ecology of coal in India*. K. Lahiri-Dutt (ed.). Aldershot: Ashgate, 219–228.

Mohai P., D. Pellow, and J.T. Roberts. 2009. Environmental justice. *Annual Review of Environmental Resources* 34: 405–430.

Nakashima, D. 2001. *Water and indigenous peoples conference report*. Geneva: United Nations.

Nash, J. 1979. *We eat the mines and the mines eat us*. New York: Columbia University Press.

Padel, F. and S. Das. 2010. Cultural genocide and the rhetoric of sustainable mining in East India. *Contemporary South Asia* 18: 333–341.

Priyadarshi, N. 2003. *The impacts of coal mining on water quality in Jharkhand*. Ranchi: Ranchi University.

Raghuvanshi, S.P., A. Chandra, and A.K. Raghav. 2006. Carbon dioxide emissions from coal based power generation in India. *Energy Conversion and Management* 47(4): 427–441.

Rao, N. 2008. *Good women do not inherit land: Politics of land and gender in India*. New Delhi: Social Science Press and Orient BlackSwan.

Rocheleau, D., B. Thomas-Slayter, and E. Wangari. 1996. Gender and environment: A feminist political ecology perspective. *Feminist political ecology: Global issues and local experiences*. D. Rocheleau, B. Thomas-Slayter, and E. Wangari (eds). London and New York: Routledge, 1–23.

Schlosberg, D. 2007. *Defining environmental justice: Theories, movements, and nature*. New York: Oxford University Press.

Scott, J. 1985. *Weapons of the weak: Everyday forms of resistance*. New York: Yale University Press.

Shiva, V. 1988. *Staying alive: Women, ecology and development*. London: Zed Books.

——. 1991. *Ecology and the politics of survival: Conflicts over natural resources in India*. New Delhi: Sage.

Singh, A.K., G.C. Mondal, S. Kumar, T.B. Singh, B.K. Tewary, and A. Sinha. 2008. Major ion chemistry, weathering processes and water quality assessment in upper catchment of Damodar River basin, India. *Environmental Geology* 54(4): 745–758.

Strang, V. 2009. *Gardening the world: Agency, identity and the ownership of water*. New York and Oxford: Berghahn Books.

Sultana, F. and A. Loftus, eds. 2012. *The right to water: Politics, governance and social struggles*. London: Routledge.

Truelove, Y. 2011. (Re-)conceptualizing water inequality in Delhi, India through a feminist political ecology framework. *Geoforum* 42(2): 143–152.

Upadhyaya, V. 2002. Water management and village groups. *Economic and Political Weekly*, December 7.

——. 2003. Customary rights over tanks: Some plain talking on limits of customs. *Economic and Political Weekly*, November 1.

Vagholikar, N., K.A. Moghe, and R. Dutta. 2003. *Undermining India: Impacts of mining on ecologically sensitive areas*. New Delhi: Kalpavriksh.

Verma, R. 2001. *Gender, land and livelihoods in east Africa: Through farmers' eyes*. Ottowa, Canada: International Development Research Centre.

Whiteford, L. and S. Whiteford. 2005. *Globalization, water & health: resource management in times of scarcity*. Oxford: James Currey Ltd.

Zwarteveen, M. and R. Meinzen-Dick. 2001. Gender and property rights in the commons: Examples of water rights in south Asia. *Agriculture and Human Values* 18: 11–25.

4 Urban water visibility in Los Angeles

Legibility and access for all

Kathleen Kambic

Water infrastructure is traditionally accessed in both public and private places. Privately, we use water in our homes and business, accessing it through sinks, water fountains, toilets, baths, laundry machines, and hoses. In public, our uses of water are more limited: we access water through public fountains (to drink and look at), and in ponds and pools. Most water infrastructure is hidden within city structures or underground, unperceived by most people thereby remaining tightly controlled and regulated.

Hidden infrastructure is an obstacle to decentralization and to the equal access to resources that are essential to human life. City infrastructures are highly centralized and regulated systems, with tight controls to limit theft and prevent security issues, to limit the physical boundary of the city, and to delineate who is in and who is not 'inside.' Water infrastructure is a highly controlled system that limits user interaction in terms of conservation, service, and accessibility through underground delivery systems and minimal community presence. This controlled water system also limits knowledge about water resources, rendering it functionally illegible to the citizenry. Thus, there is a significant mismatch in power: city entities are able to track and control water infrastructure according to agendas within existing governmental structures that may or may not reflect or even acknowledge the needs of the communities they serve.

In order to resist these power structures, I will use the city of Los Angeles to examine three responses to inaccessible water infrastructure. By situating water infrastructure issues faced by the city of Los Angeles within landscape architecture discourses and the conceptual frames of feminist political ecology and intersectionality, I make the case that without connecting urban design practices to the social and cultural paradigms of the city, it is difficult to understand who is controlling the city and through what forces this control occurs. Three main problems with control of water resources are identified, and then approaches are posited that unpack and clarify how designers, cities, and individuals might develop more equitable use of urban water resources. These three approaches—accepting the situation, incremental infrastructure, and the Cyborg City—address contemporary efforts to balance individual needs, the presence of nature in the city, and industrialization within marginalized/orphaned spaces. These three solutions make possible a responsible and responsive set of methods for addressing First World water issues in a decentralized feminist mode.

Framing water infrastructure within a feminist dialogue

Economic flows and power flows are related to water flows—people get water through four basic mechanisms: 1. hidden pipe structures where water in one's glass is literally disconnected from its sources, 2. large-scale movement of water across cities and regions, 3. systematic commoditization and privatization of all water sources (from open water access, to bottled water, lack of public water sources, to internalized delivery), and 4. private sources, such as wells or historic riparian water rights. Many people have little choice in their options to access water and have little or no recourse to change this condition, particularly where water infrastructure is unevenly distributed because of dependency on the internalized urban system. This is echoed in the increasingly mechanized, large-scale systems of infrastructure that limit our interaction with these internalized systems, as though water is 'produced' and its quality and quantity are maintained through magic.

A feminist political ecology approach for the study of water, urban design, and landscape architecture calls attention to ways in which powered interests affect urban citizens' lived experiences in public spaces along the lines of multiple axes of difference (for this study, gender, social class, and geographical location).

To understand how internalized systems restrict alternative uses of and access to water,[1] it is helpful to look at cities through the lens of political ecology. According to Swyngedouw, political ecology explains interconnected patterns of social and environmental change, where through such study, one can identify and combat inequity that can achieve "a more inclusive mode of environmental production" (2004, 24). Modernism and capitalism are two systems that regulated urban infrastructures, environmental resource management, and how the feminine was understood in the nineteenth and twentieth centuries. Modernity is obsessed with power—the power to name, classify, and categorize the world in specific ways (King 2004, 73). Patricia Morton asserts that according to feminist theorists of the 1960s and 1970s, "the sexual binary male/female constructs a series of negative values specifically related to power; the female means passivity, powerlessness, death, the natural, irrationality, and the Other, whereas the male connotes activity, power, life, the cultural, rationality, and the Self" (2010, 277). Often, such roles of masculine and feminine continue to be present in associations of water resources and scale within the city, which reinforce a binary of control in managing urban water and influence popular conceptualizations of urban water as dark, mysterious, and dirty due to its 'uncontrollable' nature.

Modernism presented the opportunity to clearly delineate large infrastructure works in the city as good, clean, new, and rational, precluding the inclusion of individual contributions and variety:

> Industrial capitalism and its social and cultural effects provide the basis for notions of the 'modern'. Its preindustrial 'predecessors', the rural, the handmade, and that which is disappearing, become the 'traditional'. These debates occur across, and help to inform, the spaces and identity of the

> nation . . . It was a discourse that emerged relationally, dialectically, and it always occurred within an unseen distribution of power.
>
> (King 2004, 72)

The Modern capitalist city is framed as part of a national collective identity, which in turn often alienates minority groups and individuals alike. Views of nature and the city were codified into a binary relationship that requires the city to remain a stable, non-natural, static entity. Intersectionality frames help us examine the interconnecting issues of gender, social class, and ethnicity (Hill Collins 2009) as they play out in power relations involved in the design of urban areas.

"Capitalism created new avenues for the development of human bodies, thoughts, and feelings, even for escaping from the existing order" (Theweleit 2010, 382), but these avenues had to be channelized, focused, and encoded to restrict the possibilities allowed under that system. To further explain what was encoded, "under patriarchy, the productive force of women has been effectively excluded from participation in the male public and social productions," which is clear in the design arts like architecture and landscape architecture (ibid.). The dominant class, often including women, used women as placeholders or showpieces within the patriarchal system. "This has enabled men to see and use them collectively as part of the earth's inorganic body – the terrain of men's own productions" (ibid., 384). The city, although feminine in many languages, was fundamentally a male construct that manifested power through control over resource infrastructures and systematization of particular binary relationships and organized streams of power. In addition, even dominant women could not affect the uses of the city and so were themselves reinforcing readings of the 'City Beautiful,' which was literally played out in public space.[2] At the turn of the nineteenth century, the City Beautiful movement epitomized the idea that cities could provide a better life to people through regulation of sanitary living and other conditions through a moralized 'civic virtue' that would protect the upper classes from disease, squalor, and even the appearance of the lower classes in public space. This movement depended upon large-scale infrastructure changes in cities, such as the construction of monumental civic centers, like the National Mall in Washington DC, which promoted the reading of that space as the embodiment of the Founding Fathers' ideals that reinforced the legitimization of the government.

Access and legibility

In the nineteenth century, waste and wastewater became anathema to proper functioning of a Modern, sanitary city. The control of water resources became paramount to prevent the spread of disease and reduce traffic congestion, among other things. As Swyngedouw (2004) contends, water's domestication through regulated large-scale systems and home delivery led to new ways to manipulate the citizenry. Water infrastructure became hidden arteries of cities, connected to transportation systems, distinguishing the classes with exclusionary tactics. Cities became consumed with scrubbing the urban surface to promote cleanliness through controlling natural elements, displacing lower-class communities, and most importantly, by removing rain and sewer water. Best embodied by Paris in the mid-nineteenth century,[3] cities built contemporary sewer systems, many with pipes that are still in operation, and obliterated poorer neighborhoods in the process. This was done partly to open up the medieval city in favor of grander boulevards that offered more light and fresh air, as well as for crowd control. In urban areas throughout industrialized countries, poor neighborhoods were seen as prime locations for redevelopment of the city through infrastructural initiatives, and particularly for the creation of underground water systems. Philadelphia, London, and Paris all raced to have the most modern (and 'invisible') sewer systems (Burian and Edwards 2011).

In contrast to the rural and agricultural land supporting the city, the urban surface does not rely on visible, accessible water to function.[4] The very sustainability of cities and the practices of everyday life that constitute 'the urban' are predicated upon and conditioned by the supply, circulation, and elimination of water. Contrary to the rural realm where—at least under non-arid conditions—water of a reasonable quality is easily and often readily available, urban water supply and access relies on the perpetual transformation, mastering, and harnessing of 'natural' water.[5] This is becoming less true, however, as rural towns in California, New Mexico, Arizona, and Texas are running out, or have run out, of water already.[6] Over 80 percent of the water in Los Angeles is imported from other watersheds (Los Angeles Water and Power). Wells and surface water sources in the West have been consistently running dry in the last few years, with the over-taxed Colorado River being the poster child for rivers no longer able to reach the sea.[7]

In order to change the symbolic and cultural meanings of water, legibility of water must increase in physical, visual, and economic ways. Presently, legibility of water in cities is dependent on its visibility, primarily operating through an understanding of where water comes from and goes. Simple efforts like stencils on gutters, visible pipes, sounds of water, water company labels, and safety signage help people to understand the limits of their watersheds and their localized water cycle. The signage identifying the extents of the Chesapeake Bay watershed is one good example, where this signage occurs along highways and other high visibility areas.[8] Legibility of water movement is also dependent on the types of access people have to clean, or any, water. It is important to remember that today people almost always have to pay for access to water: it is no longer common in most cities to see free drinking fountains, free bathrooms, or free pools.[9] In addition, municipalities are limited by economic conditions in order to be able to build and maintain water infrastructure.[10] So, if water infrastructure is creating conditions that define and require physical and economic limits to the space of the city and to residents' use of that urban space, how do cities begin to dismantle these limits and create more water legibility?

Los Angeles context

Across the country, people have been led to believe that large-scale water infrastructure, and its resultant costs, is the only way to provide clean water to citizens, farms, and other consumers. In the last 90 years in California alone, there has been a major acceleration of water projects, from the Lee's Ferry compact, to the Hoover Dam, Los Angeles aqueduct, and Salton Sea. Los Angeles in particular has been both the beneficiary and the punching bag for water profiteers. Los Angeles is in both fiscal and hydrological debt, most spectacularly encapsulated in the recent attempts of investors[11] to buy water from the Great Lakes and truck it across the country to Los Angeles and elsewhere. Some reasons people from Los Angeles are trapped in this consumerist debt cycle are that the infrastructure of Los Angeles is massive and has a life of its own, there are no places where natural or undisturbed water sources exist within the city, and the way people see water is largely regulated by the city's complex history and tense economic relationships with water and industrial production.

The Los Angeles River and the Los Angeles Aqueduct exemplify the management, regulation, and collision of knowledge and water systems, when, in the mid-nineteenth century, money started pouring into the city due to silver, lead, and zinc mining in Cerro Gordo. Water consumption was not regulated and continued unabated until late in the century when the water from the Los Angeles River and its ditch system were overtaxed, anticipating the need for a new, larger resource. The Los Angeles Aqueduct was completed in 1913 by the city under the auspices of William Mulholland. Once completed, the aqueduct became in effect the new river, and extended the city's watershed some 350 miles north to the Mono Lake Basin and Owens Valley (Lehrman 2009, 24) (see Figure 4.1). Los Angeles' consumptive watershed increased exponentially, drawing from across the state to deliver ten times the amount of water previously available to the city. Now, the current water basin for Los Angeles is subcontinental in scale and extends across eight states. Tap water in Los Angeles includes runoff from the Rocky Mountains of Wyoming and Colorado, the Sierra Nevadas, and the southern Cascades (MacDonald 2012).

This systematization of the river helped prevent flooding, but also cut off the river from its natural banks, from informal public use, and from its ecological habitat, thus marginalizing the river and redistributing its access:

> The present-day river functions mainly as a flood-control system consisting of tributary debris basins that capture sediment from the mountains, dams, and reservoirs that regulate and detain water, and a concrete riverbed engineered to conduct water to the ocean as quickly as possible . . . the river has ceased to exist as a single entity . . . Through its many reaches, the river channel is a shattered mosaic of public and private ownerships, with parcels going into and through the channel.
>
> (Fletcher 2009, 40)

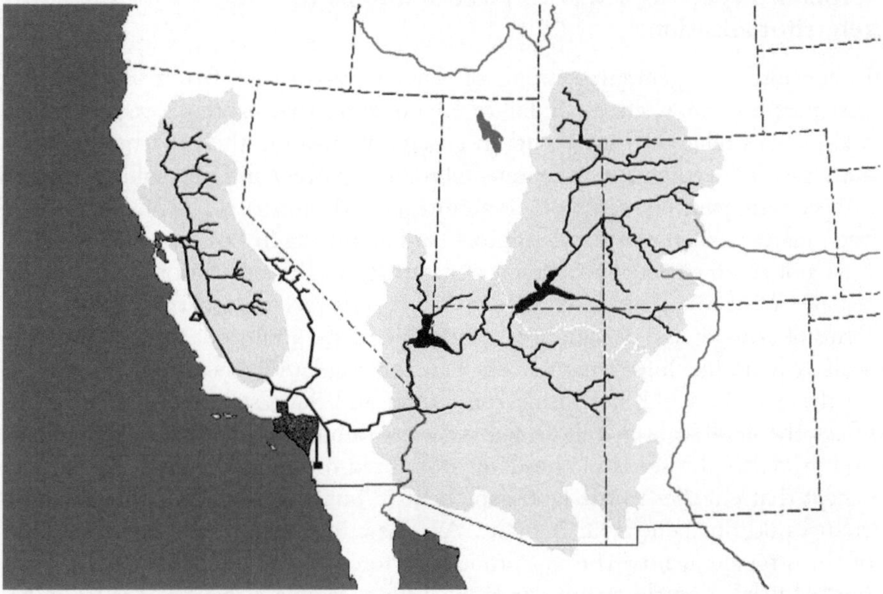

Figure 4.1 Los Angeles watershed drainage zone (shaded grey) with supply watersheds and water channels (aqueducts indicated with straight lines)

Source: Kathleen Kambic

Water in Los Angeles is very much a part of the city's persona, while simultaneously embedded and hidden within the city in large-scale infrastructural projects.

Infrastructure is a tool of reterritorialization, where "reterritorialization constructs human subjects as masters of the machine – directors and channelers of their own, and of social streams" (Theweleit 2010, 380); where, "controlling the flow of water implies controlling the city, as without the uninterrupted flowing of water, the city's metabolism would come to a halt" (Swyngedouw 2004, 2). When the Los Angeles Aqueduct retired the Los Angeles River and its ditch system from use and deprived the Owens Valley of its ability to act as an agricultural area and destroyed its natural ecology, it became a good example of how infrastructure restricted the water access of both local people in Los Angeles and distant Owens Valley farmers. In addition, the now codified delivery of water led the people of the city to believe no more floods or water shortages could occur. These exclusionary practices within the water system disempower communities and prevent opportunities to turn neighborhoods around from natural resource poor zones to thriving locales with visible natural elements that support community growth and health.

Orphaned space as a way to place solutions to reterritorialization

In contrast to hegemonic systems of water infrastructure, feminist artists and designers turn to 'women's crafts' to create works that express women's values and lives as related to water, urban environments, and the city through such activities as "needlework, weaving, embroidery, pottery and rug making, instead of becoming painters, sculptors or architects" (Morton 2010, 277). There have been many women painters, sculptors, and architects in Los Angeles, but they have not received equal acclaim as male artists for the sole reason that they are female. Women's crafts have received even less recognition than their other forms of artwork and design, or than men's designs. Crafts do not look at large-scale systems, like infrastructure. They are self-regulating arts that do not impose on the outside world, but rather reinterpret and share stories of the world. In effect, the small scale of craft-art fights the continuum of large-scale development and produces the seeds of the deterritorialized future of the city. We blindly accept that changes to the landscape benefit 'humans' when in actuality those changes do not benefit all humans. As Theweleit argues, "the concrete form of the struggle against the . . . productive force of the unconscious has been (and still is) a battle against women, against female sexuality" (2010, 376). This battle is waged in the marginalized, non-capitalist spaces resistant to the Modern, also known by interior designers as orphaned space. Similar to the marginalized activities of (women's) crafts that are often overlooked, orphaned spaces are the leftovers, the marginalized zones that are engulfed by the male dominated infrastructural, construction, and celebratory spaces of the city. The materials these orphaned spaces are created with, or re-appropriated through, are often leftovers, recycled, salvaged, or old. Their designs, whether space for refuge, gardening, or other uses, are homegrown, that is, created by the individual without design training. Though their functions might serve the citizenry at large, for instance as a garden, their construction is often done by a single individual.

According to Charles Mudede, "[o]rphaned spaces include so-called public wilderness, 'sites that are within the city, but not part of it, and so confuse (or worry) the line between in and out, between country and urban, between natural and social'" (Mudede 2010). Urban rivers are commonly orphaned spaces because they connote danger, uncleanliness, mystery, the wild, the feminine. Other such spaces are underpasses, vacant lots, highway medians, even local parks or individual yards. Orphaned space is an opportunity for people to reengage the city untethered from traditional structural limitations while also creating space where the feminine is celebrated and given opportunity to express itself.

Orphaned space is considered part of the system of landscape in the city in contemporary design practice, as evidenced by the Landscape Urbanism movement.[12] But how do we define 'landscape' outside of this particular practice? Two comments J.B. Jackson makes in his book, *Discovering the vernacular landscape*, are notable: 1. that "the ceaseless interaction between the ephemeral, the mobile, the vernacular on the one hand and the authority of legally established premeditated permanent forms on the other is of primary importance" and 2. that landscape is "an environment where permanence and change have struck a balance" (Jackson 1986, 148).[13] Mobility and dynamic equilibrium are the calling cards of urban landscape that evoke the feminine in terms of flexibility, lack of formality, and mutability.

Orphaned space appears in three paradigms of thought: the first paradigm is 'cosmic shift,' where landscape went from being actual to metaphorical. Humans shifted the landscape from the fundamental place of occupation to a secondary representation of 'outside,' i.e. the nature vs. culture dichotomy. The manifestation of landscape in the city was, by nature, feminine or other marginalized space—it was the advent of the garden and of orphaned space.[14]

The second paradigm is the commodification of water and landscape elements: use of natural systems shifted to man-made ones. Increased scale of farming required advancements in water delivery, as yields were required to increase yearly, rather than remain stable. The commodification of water, or the relationship between water and energy, emerged as a new power dynamic in cities as economic pressures ran up against limits of availability and access. Two main issues with the commodification of water are limited distribution and capacity. Water, although renewable, is not instantaneously so; we are using it faster than it can be recycled through the natural water cycle.[15] These issues are not commonly known or discussed, partially because water infrastructure (especially sewage and industrial treatment) is invisible or relegated to edges of populated areas, constituting urban scale orphaned space.

The third paradigm is the primary fit of humans to the landscape: once humans fit themselves into the land, they orient themselves to it (Barlow Rogers 2001). Infrastructure was created to modify land to people's needs and increase production at the scale of the city, not of the individual. The increase in scale of infrastructure has created new problems politically, socially, and economically, reinforcing the idea of the city as the "terrain of men's own productions" as noted above. For instance, 240,000 water main breaks occur each year in the United States, according to the Environmental Protection Agency's (EPA) Aging Water Infrastructure Research Program. Some pipes are up to 200 years old, many are made of outdated materials like terra cotta, and most are taxed to carry higher pressures than originally intended (U.S. EPA 2002). Water is a limited resource; building more dams won't hold water that doesn't fall from the sky, which means that the struggle between growing metropolises and water supplies will only intensify. Los Angeles and Las Vegas are two of the most powerful water interest groups, and knowing that, at the end of the day, like most things, "water flows uphill toward money," will shift the power balance of those with water and those who need it (Gertner 2007, 21). "[T]he political ecological history of many cities . . . written from the perspective of the need to urbanize and domesticate nature's water and the parallel necessity to push the ecological frontier outward as the city expand[s]" (Swyngedouw 2004, 37), promotes ex-urban and sub-urban growth and urban consumption, which are unsustainable economically and ecologically. It also denies recognition of its bi-product, that of orphaned space.

Shifting to deterritorialization occurs in three ways

Each of these paradigms can be shifted through new perspectives appearing in contemporary landscape architectural discourse.

1. Accepting the situation addresses COSMIC SHIFT

Accepting the situation is about designing in conjunction with the existing condition, not an idealized past or future. Presently in cities, the entire urban surface is managed in some way or another, whether as a productive or orphaned space, through controls, rules, and restrictions set by governmental agencies. Peter Latz, designer of the Duisburg Nord landscape park in Germany, built on former industrial land but now accessible to all as a recreational park for all ages, asserts the idea of accepting the situation as a new direction for design. According to Rosenberg,

> Latz's notion of "accepting the situation" is not a passive one, but just the opposite; it is at the root of his inventiveness. His acceptance of the existing reality impels him, on the one hand, to discover and exploit [the site's] unique qualities in unusual ways; and on the other hand, to search for new solutions to the problems it poses.
>
> (2009, 216)

Latz's approach is empirical, physical, and experimental, engaging all the materials of the industrial site, such as retaining part of the old plants with their old machines. Rosenberg quotes Latz, "Everything is *good*, even polluted soils (except toxic ones), since everything is recycled." Soils were improved through the use of plants to uptake toxic substances through a process called phytoremediation. She continues,

> Accepting the situation is about being at home . . . in the physical, material world. It is about *making*, more than *healing* . . . it is about living with contradictions in a fragmentary world, and abandoning any notion of an idealized nature, apart from man.
>
> (Ibid.)

Rosenberg's assertion that everyday life is primary when we speak about nature and things that are natural derives from the notion that humans need to have daily contact with nature. In Los Angeles, the river was once a natural phenomenon that urban residents and farmers depended upon. It could once again be this thing, but only if we allow the existing and voluntary hybrid ecologies to occur in conjunction with existing infrastructure, such as highways, orphaned spaces, industrial areas, etc.

A response in this vein would be more economically viable than the present proposed re-design of the Los Angeles River, by the LA River Corporation, Mia Lehrer and Associates, and many others. The 'revitalization' would completely re-destroy an operating ecological system to rebuild the river corridor into some sort of idealized neo-liberal space. The redevelopment projects will certainly remove any traces of the recent past of the river and render the informal communities of people, animals, and plants ineffective and displaced.

2. *Small-scale intervention addresses* PRODUCTION

The best example of deterritorialization is the incremental infrastructure of small-scale interventions, which occur in three forms: daylighting, catchment systems (or rainwater harvesting), and urban agriculture. Daylighting is the act of bringing a formerly piped stream or rivulet out of pipes and back to the surface of the ground. Daylighting and constructing small-scale water catchment systems could be understood as acts of revealing the subconscious. Decentralizing infrastructure could be seen as a feminist guerilla act, evidenced by work done by the Greywater Corps in Los Angeles.[16] Although this work is very 'homegrown,' it is notable that people using these systems are able to repeatedly reuse water from the urban water supply to support themselves, their families, and yards, through, for example, small-scale agriculture, gardening, water reuse, and greywater storage.[17] Constructing new methods of interacting with city structures, like sidewalks, roofs, and open lots, enables people to employ all types of skills, especially those that are not part of a codified engineering education, to improve their access to water and their ability to utilize that water for production, and to improve their local neighborhood aesthetically.

The water harvesting systems that the Greywater Corps (and similar groups throughout the country) employ are small-scale and easily constructed. This type of development recalls the self-regulating, decentralized 'women's crafts' of the turn of the twentieth century. In fact, many of these companies are run by women, who are often not called 'entrepreneurs,' but 'hobbyists' (Sauers 2013). Greywater Corps has noted that,

> In 2010 the State of California rewrote decades of restrictive building codes to allow simple residential greywater reuse. Our systems typically save a household 10,000 to 50,000 gallons per year – water that would otherwise be pumped hundreds of miles from the Colorado or Sacramento rivers at a huge material and ecological cost.
>
> (Greywater Corps 2014)

Projects include branch drain systems, greywater pipes from interior uses to landscape zones, and moveable pipe systems, all of which can be built using low-cost materials and installed in short order. Efforts like these are incrementally preventing Los Angeles and other cities from becoming victims of large-scale infrastructural failure.

Urban agriculture is another common method of deterritorialization in cities. Whether a return of the Victory Garden, a necessity due to economic circumstance, or part of the Slow Food movement, urban agriculture is filling the role of ecological production in cities. Without sustainable cities, it is difficult to imagine a world where natural systems are in balance with human development, and food and energy resources are secure.

> Urban agglomerations and their resource uses are becoming the dominant feature of the human presence on earth, profoundly changing humanity's relationship to its host planet and its ecosystems . . . The cities of the 21st century are where the future of the biosphere will be played out.
>
> (Deelstra and Girardet 2000, 43)

In the most incremental manner, individuals are able to reduce their dependence on the dominant infrastructure of the agricultural machine and, if they practice passive or active rainwater harvesting, on municipal water, to produce goods for their discreet communities.

In Los Angeles, urban agriculture ranges from small garden plots, to rooftops, to larger empty lot farms. Farmscape, a local start-up, has been struggling with water availability, soil contaminants, biodiversity, and political difficulties.[18] They proactively maintain the gardens for efficiency and health, and provide education for the community. Los Angeles is the urban farming capital of the United States, based on the year-round growing season, ample sunlight, and good soil. Los Angeles is home to the South Central Farmers, a group of urban, mainly Hispanic, farmers in south Los Angeles that have inspired the nation partly through a film, *The Garden*,[19] that chronicled their successful transformation of a barren and blighted 14-acre area in Los Angeles into a verdant, food-producing garden that fostered inter-generational community building. Despite their controversial eviction from their land in 2006 for private redevelopment due to city, business, and racially oriented politics, some of the urban farmers from this group have continued to grow fresh produce in Bakersfield for community supported agriculture (CSA) distribution while others joined the Stanford-Avalon Community Garden.[20] Through programs of urban gardening and rain harvesting, Los Angeles is leading the way for communities and individuals to separate themselves from municipal infrastructures of water and food with low-cost, recyclable, and easily maintained systems.

3. *Cyborg City addresses* PRIMARY FIT TO LANDSCAPE

Elmhirst (2011) explains that feminist political ecology (FPE) is expanding in terms of publications on the relationship between gender and resource access, specifically where land allocation, the deconstruction of the commons, and governance are involved. In this study of Los Angeles, access to water is examined from a gender and social class perspective and the commons can be viewed as synonymous with urban orphaned spaces such as rivers and parks. Elmhirst explains that work in FPE connects gender, space, race, and other signifiers to expand understanding of 'the intimate' and how it relates directly to power relations at the scale of the city and beyond. Another way to address the inequities and environmental harm stemming from masculine water infrastructure/political ecology of the city is the Cyborg City concept. As an offshoot of 'accepting the situation,' the Cyborg City accepts input from both humans and natural systems to reorient our 'primary fit to the landscape.' This is because entities like the Los Angeles River are no longer 'only' natural. They include vehicular infrastructure like bridges and viaducts, power lines, pathways, fences, and sewers. Haraway defines it well when she states, "a cyborg world might be about lived social and bodily realities in which people are not afraid of their joint kinship with animals and machines, not afraid of permanently partial identities and contradictory standpoints" (2013, 154). From a landscape architecture standpoint, Cyborg Cities also contain governmental structures like right-of-ways and easements, regulated uses for commerce like docks, and cultural structures like gang hang-outs or picnic areas. Fletcher states that "these fluid systems are more evident in the political and social operations than physical form: rivers of energy, streams of revenue and resources, movement of goods and services" (2009, 36–38) that flow amongst each other, not in discreet zones.

Infrastructure and ecology can become a singular entity of mutual reinforcement, especially in orphaned spaces. However, most often in Los Angeles, 'disasters' such as mudslides and floods are the result of developers building in precarious environmental places in which nobody should build. Attempts by governmental entities to remedy these conditions or, alternatively, to dream of restoring the landscape, turn to legislation or to complex infrastructural systems for some degree of security. This can be seen most recently in response to Hurricane Sandy in New Jersey and New York, where neighborhoods on Staten Island are being returned to wetland, displacing many communities and turning them into 'buyout zones' whereby the government purchases land most at risk from flooding (Gregory 2013). Terrain is reduced to controllable, ordered plots of land with specific characteristics and descriptive boundaries, efforts that actually compound rather than simplify the socio-ecological condition of the urban landscape today.

Opportunities already exist to design for both social and ecological needs in Los Angeles. As the urban designer and landscape architect, David Fletcher, has argued, "this is an infrastructural ecology, opportunistic and emergent, one that lives off human excess, with many of its values and functions unknown or misunderstood" (2009, 36–38). Unable to operate in their traditional ecological manner, rivers, estuaries, and wetlands in cities are often considered broken, missing, or damaged by the citizenry. If these voluntary hybrid ecologies could promote other infrastructures existing within the river corridor, the space might be seen as a community asset rather than a land of "illicit boundaries, with gang territories crisscrossing its pathways and spaces, and human encampments within storm drains and under bridges" (ibid., 40). Current users of the space along the river should be recognized in re-design efforts; an intersectionality framework can remind us of the varied social groups inhabiting the river's spaces and the power dynamics involved in re-design efforts. We should remember that even if the river does not operate in a known ecological paradigm, it does not mean that it is not operating at all. In fact, the river could be designed to play larger roles, or better fit the roles it presently embodies, and have many different symbiotic functions. At present, 11 miles of natural river habitat have been restored; the Los Angeles River Center and Gardens provide green space and an exhibit on the river restoration project that includes future plans and is part of a river trail with several pocket parks with native vegetation including in areas where residents' lives were previously dominated by highways and dense urban development (Pulido et al. 2012).

Two images of the Los Angeles River are presented here: one from a site in Glendale, CA originally developed by the Army Corps of Engineers, and one developed by the City of Los Angeles for the cover of the Los Angeles River Revitalization Master Plan, in conjunction with many consultants, community groups, and government agencies. From these images, it is possible to see the similarities and differences between them (see Figures 4.2 and 4.3). Both sites were developed primarily for flood control.

These two images clearly demonstrate the difference between the Cyborg City and contemporary design practices that do not 'accept the situation.' First looking at the image of Glendale, it is important to note that this is a stretch of the river

Figure 4.2 Los Angeles River at Glendale Narrows
Photo credit: Kathleen Kambic

Figure 4.3 Los Angeles River Revitalization Master Plan

Source: City of Los Angeles. 2007. Los Angeles River revitalization master plan. Department of Public Works, Bureau of Engineering. www.lariver.org

that does not have a concrete bottom to its culvert. This allows for volunteer species to more easily adapt to the site's conditions because soil is already available. As one can see, trees have established themselves in this area and the movement of water around these small islands is slowly reestablishing sinuosity into the water flow.[21]

In the second image—the rendered drawing of the new and improved Los Angeles River—the situation is slightly different. Instead of allowing for the shifting course of the river to change over time, the walkway to the right permanently sets its path through the large concrete steps (and perhaps gabion walls). Also, the image shows an idealized perspective of the relationship between the rocks, grasses, and low shrubs that might inhabit the river bed itself, where they are built up in a pattern similar to a planting bed with relatively even distributions. Lastly, the trees on both sides are cut off from the water channel itself. The trees are likely to be irrigated through a designed system set into the beds in which they are planted, completely cutting them off from the ecology of the river. Thus, while Figure 4.2 presents us with a somewhat disorganized agglomeration of trees within the river culvert corridor, it is a more highly functioning system than in Figure 4.3, where the natural systems have been made as idealized versions of themselves that do not actually function ecologically and are highly controlled spaces reflecting ideals of both the government and the designers, over nature and chance.

As both a teacher and practitioner of landscape architecture, Fletcher has elucidated, "[t]he naïve desire to return the river to a 'natural' state amidst an asphalt metropolis is, in fact, a threat to the urban ecologies that have emerged in response to the river's modifications" (2009, 46). The resultant hybrid systems, whether we call them green infrastructure or accidental nature, process excess nutrients, control erosion, and provide habitat for urban species, independent of human agency. These places are wild (or feminine) in every sense: new visions of the Cyborg City can be projected onto them, or they can be left intact to develop on their own. They shift over time, where the time scale is larger than human city building projects. These hybrid systems appear incrementally and in overlapping phases, where no specific end point is trying to be reached. Where hegemonic systems are often shortsighted and limited in scope to necessitate a specific outcome, the Cyborg City promotes variety, incrementalism, and potential long-term health benefits and accessibility for all citizens. These hybrid systems exist at multiple scales, the domestic small-scale intervention of the city plot, the large scale of the river itself, and even regionally, in terms of the total watershed of Los Angeles.

Conclusion

Consequences of city-building activities include diminishing resource availability and ecological efficacy, loss of direct connection to natural systems, and altered environmental states. Fundamentally, we must come to terms with the idea that we are agents of natural processes and the processes we put in motion are a part of nature. "Stability in nature is an illusion; moreover, non-natural factors such as urbanization, global warming, and the heat-island effect all have to be included in the ecological equation" (Fletcher 2009, 46). Cities are entities that operate in tandem with nature—each ebbing and flowing against the other. Understanding that water and urban systems can act in tandem will help dismantle the hegemonic drive of infrastructural development. This chapter shows that the cyborg nature of the city can act as both site for activation of new human agency, as in the development of new smaller-scale infrastructural systems to literally dismantle and reduce dependency on the traditional hegemonic infrastructure of the city, as well as a site for natural agency, where animal and plant systems can reinsert themselves into large-scale infrastructure in ways that reintroduce the wild into the city. As King (2004) argues, the value of positively reinforcing small-scale infrastructure, personal differences and perspectives, and ecological hybridism are built upon a feminist framework. If we were to put forth a fresh paradigm that values feminist perspectives equally or more so than the existing patriarchal system that also excludes minority groups, cities and the lives of the citizenry could improve, perhaps dramatically. Intersectionality frames help reveal these connections between gender, class, and ethnicity, and relationships of power within which urban spaces and water infrastructures are created and contested. This framework therefore can also assist us in envisioning a greater variety in points of access that could be developed in the home and running through the city to allow for new readings and uses of the city.[22] Feminist political ecology frames can help to study urban power dynamics that reveal that these acts of deterritorialization enrich orphaned urban space and disenfranchised people, dismantling traditional patriarchal, hegemonic infrastructural systems to allow new possibilities for a broader spectrum of social, cultural, and ecological needs to be met, simultaneously. This incremental, site-specific infrastructure can address emerging infrastructural ecologies where systematized streams of water are equally accessed by all.

Notes

1 In Los Angeles, 68 percent of water is used for residential applications, but that water can only be accessed from a tap or hose (City of Los Angeles 2008).

2 Central Park in New York is another good example of this, where the park was originally designed as a showcase for the wealthy of the city with its promenades and allees. See Rose (1996) for information on the economic and social drivers behind the City Beautiful movement.

3 During the reign of Napoleon III, Haussmann was charged with rebuilding Paris to clean it up and provide greater crowd control to control the citizenry. The result was the boulevard and sewer systems, which ripped open the medieval fabric of Paris and replaced it with broad streets and sidewalks, fewer slums and state of the art drainage. The sewers were so popular that tours were given for the upper class to see the marvels of technology beneath their city.

4 Ironically, it is urban dwellers who support subsidies for agricultural water supply through the US Bureau of Reclamation, the Corps of Engineers and the California State Water Project.

5 In the case of Midwest farmers, availability is dependent on geography; farms west of the 100th meridian received less than 20 inches of rain per year, causing overdrafts of aquifers and river resources. See Historical background of water rights in South Dakota, http://denr.sd.gov/des/wr/history.aspx. Accessed February 6, 2013.

6 See Lovett (2013), Goldenberg (2013), and Cart (2013) for links to articles discussing the decline in water availability in these states.

7 Recently, the Colorado River reached the Gulf of California for the first time in 16 years in a pulse flow that lasted a few months, through a new agreement between Mexico and the United States. See Postel (2014).

8 Although I have seen these signs in person, there are few on any given road. Presently, the signage does not seem to have a strong effect on conservation efforts by the general public. More information on the Chesapeake Bay Foundation is available at http://cbf.org. Accessed October 12, 2013.

9 Free is a relative term in this case. The water for these amenities is commonly paid for through tax revenue.

10 See Berger (2007, 170–171) for a brief discussion on the amount of infrastructure in the United States and his Introduction for a frame of the issue of sprawl and consumption of land.

11 Different sources discuss movement of water from the Great Lakes to the Southwest and California. This information was related to the author through personal anecdote. See Jones (2007).

12 According to James Corner, landscape is a series of "systems [that] condition the distribution and density of urban form" (2006, 28). The urban is defined as a surface upon which actions and actors play out the urban condition.

13 Jackson states,

> It may be that I am here on the track of that elusive landscape concept: the ideal landscape defined not as static utopia dedicated to ecological or social or religious principles, but as an environment where permanence and change have struck a balance.
>
> (1986, 148)

See Chapter 13, "Concluding with Landscapes," for a detailed discussion of types of landscapes and their definitions.

14 Images created for the Duc du Berry, near Paris, in the book *Très Riches Heures* (1416) by the Limbourg brothers, Herman, Paul, and Johan (painters), richly illustrate this idea.

15 New York City completely recycles its water supply every 24 hours, to extinction. See Kaufman (2008). In addition, disruption of other cycles, like nitrogen, carbon, and salts, and shifts in temperature, can even more quickly exhaust water supplies.

16 The Greywater Guerrillas in Seattle are a similar group of repute. Their book, *DamNation* (Wolfe-Erskine, Cole, and Allen 2007) along with works by Brad Lancaster who is based in Tucson, Arizona and author of *Rainwater harvesting for drylands and beyond* Volumes I and II, are good sources for individuals interested in catchment systems (Lancaster 2006).

17 A typical greywater system entails gathering water off hard surfaces during a storm event, collecting and filtering that water, later using the water inside the house for some purpose and then returning that water to the ground for infiltration through plant beds. Other typical systems use municipal water that was supplied to a home or other building, used in it, and then channeled for reuse in landscaping. These systems clearly decentralize the currently highly centralized system of wastewater treatment in both the United States and Australia, as presented in Brad Lancaster's (2006) writing and Jerry Yudelson's books, *The world's greenest buildings* (Yudelson and Meyer 2013) and *Dry run: Preventing the next urban water crisis* (Yudelson 2013), which detail these systems in both domestic and commercial buildings, including discussions of their efficacy.

18 See http://farmscapegardens.com. Accessed March 2, 2014.

19 See *The Garden*, www.thegardenmovie.com. Accessed March 2, 2014.

20 See "Los Angeles is the urban farming capital of the U.S." Farmscape, January 11, 2012, http://farmscapegardens.com/blog/los-angeles-urban-farming-capital-us. Accessed March 2, 2014.

21 This can be seen in the ripples reflected in the surface of the water. Areas that have more reflection occur where the bottom of the channel is shallower, where the smoother areas demonstrate that pools are starting to develop. This balance between riffle and pool creates a naturalized condition that supports an increased amount of biota overall. See writings by Forman and Godron (1986) for further information.

22 See Morrish (1984) for information on the development of the Los Angeles supply watershed and ideas to reveal the watershed structures within the city.

References

Barlow Rogers, E. 2001. *Landscape design: A cultural and architectural history*. New York: Harry N. Abrams.

Berger, A. 2007. *Drosscape: Wasting land in urban America*. New York: Princeton Architectural Press.

Burian, S.J. and F.G. Edwards. 2011. Historical perspectives of urban drainage. In *Politics of urban runoff: Nature, technology, and the sustainable city*, ed. A. Karvonen, 1–16. Cambridge, MA: MIT Press.

Cart, J. 2013. New Mexico is the driest of the dry. *Los Angeles Times*, August 6. www.latimes.com/news/nationworld/nation/la-me-parched-20130806-dto,0,5922502.htmlstory. Accessed December 15, 2013.

City of Los Angeles. 2007. Los Angeles River revitalization master plan. Department of Public Works, Bureau of Engineering. www.lariver.org. Accessed March 30, 2013.

——. 2008. Securing L.A.'s water supply. Department of Water and Power (LADWP), December 1. www.lacity.org/mayor/stellent/groups/electedofficials/@myr_ch_contributor/documents/contributor_web_content/lacity_004714.pdf. Accessed March 30, 2013.

Corner, J. 2006. Terra Fluxus. In *The Landscape Urbanism Reader*, ed. C. Waldheim, 21–34. New York: Princeton Architectural Press.

Deelstra, T. and H. Girardet. 2000. Urban agriculture and sustainable cities. In *Growing cities, growing food: Urban agriculture on the policy agenda*, eds N. Bakker, M. Dubbeling, S. Gündel, U. Sabel-Koschella, and H. de Zeeuw, 43–65. Feldafing, Germany: Deutsche stiftung für internationale Entwicklung.

Elmhirst, R. 2011. Introducing new feminist political ecologies. In *Geoforum* 42(2): 129–132. doi: 10.1016/j.geoforum.2011.01.006.

Fletcher, D. 2009, Flood control freakology: Los Angeles River watershed. In *The infrastructural city: Networked ecologies in Los Angeles*, ed. K. Varnelis, 34–51. Barcelona: Actar.

Forman, R.T.T. and M. Godron. 1986. *Landscape ecology*. Hoboken, NJ: Wiley.

Gertner, J. 2007. The future is drying up. *New York Times*, October 21.

Goldenberg, S. 2013. A Texan tragedy: Ample oil, no water. *Mother Jones*, August. www.motherjones.com/environment/2013/08/fracking-texas-drought-climate-change. Accessed December 15, 2013.

Gregory, K. 2013. Deciding whether it's lights out. *New York Times*, N.Y./Region. October 25.

Greywater Corps. 2014. About Greywater. http://greywatercorps.com/about01.html. Accessed June 11, 2014.

Haraway, D. 2013. A Cyborg manifesto: Science, technology, and socialist-feminism in the late twentieth century. In *Simians, cyborgs and women: The reinvention of nature*, ed. D. Haraway, 149–181. New York: Routledge.

Hill Collins, P. 2009. *Emerging intersections: Race, class, and gender in theory, policy, and practice*, eds B. Thornton Dill and R. Zambrana, vii–xiv. New Brunswick, NJ: Rutgers University Press.

Jackson, J.B. 1986. *Discovering the vernacular landscape*. New Haven, CT: Yale University Press.

Jones, T. 2007. Fate of Great Lakes' water looking fluid. *Los Angeles Times*, Water Shortages, October 28. http://articles.latimes.com/2007/oct/28/nation/na-water28. Accessed March 20, 2013.

Kaufman, F. 2008. Wasteland: A journey through America's cloaca. *Harper's Magazine*, February, 46–59.

King, A.D. 2004. *Spaces of global cultures: Architecture, urbanism, identity*. London: Routledge.

Lancaster, B. 2006. *Rainwater harvesting for drylands and beyond*. Tucson, AZ: Rainsource Press.

Lehrman, B. 2009. Reconstructing the void: Owens Lake. In *The infrastructural city: Networked ecologies in Los Angeles*, ed. K. Varnelis, 20–33. Barcelona: Actar.

Lovett, I. 2013. Arid southwest cities' plea: Lose the lawn. *New York Times*, August 12. www.nytimes.com/2013/08/12/us/to-save-water-parched-southwest-cities-ask-homeowners-to-lose-their-lawns.html?pagewanted=all. Accessed December 8, 2014.

MacDonald, G.A. 2012. Los Angeles water – Myths, miracles, mayhem and William Mulholland. AAG *Newsletter*, December 5. www.aag.org/cs/news_detail?pressrelease.id=1612. Accessed January 20, 2014.

Morrish, W.R. 1984. The urban spring: Formalizing the water system of Los Angeles. In *Modulus 17: The University of Virginia architectural review*, 44–73. Charlottesville, VA: The University of Virginia School of Architecture.

Morton, P. 2010. The social and the poetic: Feminist practices in architecture, 1970–2000. In *The Feminism and visual culture reader*, ed. A. Jones, 277–281. 2nd ed. London: Routledge.

Mudede, C., ed. 2010. The politics and spaces of public wilderness. *ARCADE Magazine*. http://arcadenw.org/events/charles-mudede-the-politics-and-spaces-of-public-wilderness. Accessed January 20, 2014.

Postel, S. 2014. *A sacred reunion: The Colorado River returns to the sea*. National Geographic's Freshwater Initiative in Water Currents, May 19. http://newswatch.nationalgeographic.com/2014/05/19/a-sacred-reunion-the-colorado-river-returns-to-the-sea/. Accessed May 22, 2014.

Pulido, L., L.R. Barraclough, and W. Cheng. 2012. *A people's guide to Los Angeles*. Los Angeles, CA: University of California Press.

Rose, J.K. 1996. City beautiful: The 1901 plan for Washington D.C. http://xroads.virginia.edu/~cap/citybeautiful/city.html. Accessed January 11, 2013.

Rosenberg, E. 2009. Gardens, landscape, nature: Duisburg-Nord, Germany. In *The hand and the soul: Aesthetics and ethics in architecture and art*, ed. S. Iliescu, 209–230. Charlottesville, VA: University of Virginia Press.

Sauers, J. 2013. Why are women DIY entrepreneurs called 'hobbyists'? *Jezebel*, May 5. http://jezebel.com/why-are-women-diy-entrepreneurs-called-hobbyists-494363393. Accessed May 9, 2013.

Swyngedouw, E. 2004. *Social power and the urbanization of water: Flows of power*. Oxford: Oxford University Press.

Theweleit, K. 2010. Stream/all that flows and woman: Territory of desire. In *The feminism and visual culture reader*, ed. A. Jones, 375–386. London: Routledge.

U.S. EPA. 2002. The clean water and drinking water infrastructure gap analysis. www.epa.gov/ogwdw/gapreport.pdf. Accessed June 20, 2014.

Wolfe-Erskine, C., J.O. Cole, and L. Allen. 2007. *DamNation: Dispatches from the water underground*. Brooklyn, NY: Soft Skull Press.

Yudelson, J. 2013. *Dry run: Preventing the next urban water crisis*. Gabriola Island, BC: New Society Publishers.

—— and U. Meyer. 2013. *The world's greenest buildings*. London: Routledge.

5 Advances and setbacks in women's participation in water management in Brazil

Andrea Ferreira Jacques de Moraes

Introduction: the women and water management paradox

The increased participation of women as stewards of water resources is crucial to any progress towards more responsible and innovative water management in poor rural regions of the developing world (Harris and Gantt 2007; Kevany et al. 2013). While there have been advances that acknowledge the need for gender equality in water management within the international development discourse (U.N. 2008), the policy practice of including and empowering women lags far behind (House 2003; Seager 2008; Wallace and Coles 2005; Zwarteveen and Bennett 2005). This has crucial consequences: whenever policies and programs ignore the roles of women in irrigation and water management, they render women's roles and needs invisible.

In poor rural areas of the developing world where water provision remains largely an individual or household responsibility, the management of water relies mostly on women and girls (Crow 2001; van Wijk-Sijbesma 1998). This includes fetching water from wells or ponds, managing drinking water, cooking, washing, cleaning, plus caring for family members (babies, children, the elderly, and the sick), as well as being responsible for health and hygiene. Women do these tasks in addition to working in agriculture, managing gardens, and caring for animals. Poor rural women in the developing world manage water for production and social reproduction.

Despite being the main providers of water in the household, holding essential roles in preserving water sources, being important food producers, and having primary responsibility for household health, women have less access to resources (land, credit, agricultural input, research and innovation), training and rural extension services, and to varied sources of income—all impacting their capabilities as effective water managers.

This paradox—the importance of women in water management on the one hand, and their lack of resources and power to appropriately manage water on the other hand—is often reinforced by policies and development projects that do not consider the different needs and interests of rural women as opposed to those of their male counterparts (Buechler 2005).[1] It is not surprising that even though women represent half of the world's population, they far outnumber men in terms of poverty (U.N. Women 2011). While the need to incorporate women in the water sector is acknowledged in theory, direct participation of poor rural women in water projects is often limited.

In Latin America, water policy affecting household use is often associated with women and welfare (social reproduction), while "productive" uses of water (such as the use of water for irrigation) are associated with men (Zwarteveen and Bennett 2005). However, many women are responsible for farming (and irrigation), either alongside their male partners, alone as heads of their families, or when their male partners migrate in search of temporary paid work elsewhere.[2] Still, women's work in agriculture is frequently considered as "help," and is not fully recognized and valued either by supporting agencies or by the women's own communities.

When government agencies assume that only men are farmers, gender differences in knowledge, resources, and needs regarding water irrigation are often ignored,[3] and as a consequence women are excluded from key funding opportunities and rights, such as credit and technical assistance. This, in turn, reinforces expectations that only men should participate in irrigation and water associations or other community meetings where decisions about water are made.[4] Membership in associations is often linked to titles of land and water, which are concentrated in men's hands. "Water control means political power" (Rap et al. 1999 in Zwarteveen and Bennett 2005, 24), which is traditionally associated with masculinity (Parpart 2009). If women lack representation and voice in decision-making about irrigation and water management, their priorities are also made invisible and their claims and rights are not recognized.

Inspired by a feminist political ecology approach (Buechler 2005; Elmhirst 2011; Rocheleau et al. 1996), this chapter will look at two spaces of women's participation in water management in the Brazilian semi-arid region, and at their intersection. It especially addresses the issue of scaling up (Elmhirst 2011) in water management, from the household, to the community, and to increasing levels of decision-making on water management at the territorial (micro-regional) scale. The first section describes the One Million Cisterns program in the Brazilian semi-arid region, and the program's challenges and advances related to poor rural women's participation. The second section focuses on women's participation in the National Council of Water Resources and Watershed Committees, which are territorial and participatory decision-making bodies concerned with water management. The third section considers the possibility of women participating in both programs, and the barriers and challenges experienced by poor rural women (and men). The case of the Lower São Francisco River Watershed is examined to illustrate these issues regarding program participation. Finally, a feminist political ecology perspective is utilized to address women's paradox in water—not only to shed light on the challenges faced by women at different scales of participation in water management, but also to bring to the surface the complexity of intersectionality (McCall 2005) for analyzing gender and water management.

Theory and methods

Motivated by the need to understand and address the women and water paradox, this chapter is built on three basic premises. First, it recognizes the central role of women for responsible water management (Kevany et al. 2013). It does so by adopting a feminist political ecology approach, where gender is defined as "a critical variable in shaping resource access and control, interacting with class, caste, race, culture and ethnicity" (Rocheleau et al. 1996 in Elmhirst 2011, 129). It explores the intersection of race, class, and gender (Collins 1998; Crenshaw 1991; McCall 2005; Roschelle et al. 2004) in the case of poor women in the semi-arid region of Brazil, implying that there are affinities (Haraway 1991 in Rocheleau 1995) in women's experiences with water management. How do poor rural women from the Brazilian semi-arid region participate in water management? An intra-categorical approach to complexity is sought by using scales as a way to "analytically unravel one by one the influences of gender, race, class, and so on" (McCall 2005, 1787).

Second, this chapter focuses on the different levels (or scales) of participation of poor rural women in water management, and as such, aims at contributing to "understanding the production of and interconnections between scales of analysis" (Truelove 2011, 144) from "the intimate to the global" (Elmhirst 2011, 131) or in this case from the individual and the household to the watershed scale. It asks two questions: at which levels or scales do women participate in water management? And, what are the advances, barriers, and challenges faced by poor rural women with regards to scaling up and increasing their decision-making power in water management?

To illustrate how poor rural women from the Brazilian semi-arid region participate in water management, I examine the One Million Cisterns program in the Brazilian semi-arid region and in the National Watershed Management System, and the interconnection of these programs. A mix of qualitative methods was deployed to collect and analyze data, such as interviews, analysis of primary documents, and participant observation. Interviews were conducted in 2008 with women cistern builders and other members of the *Articulação no Semi-Árido* (Articulation of the Semi-Arid, ASA) non-governmental organization (NGO) network in Afogados da Ingazeira, Mossoró, and Fortaleza, and in 2013 with government representatives at the federal level involved with ASA and the Watershed Committees. A series of government and academic documents were consulted. This included not only the National Watershed Management System, but also the specific case of the São Francisco River Watershed Committee. Quantitative data (statistics) were also collected in order to illustrate gender differences and gender participation in the National Council of Water Resources, "as evidence for the pervasiveness and distribution of the problem [of gender inequalities in water management], not just the nature or seriousness of it" (Rocheleau 1995, 461).

Women's participation in the One Million Cisterns program[5]

Brazil has one of the most participatory water management laws in the world. This means that in legal terms a poor rural woman could be a civil society member of the Watershed Committee representing her community. As a member of the Watershed Committee she could help to decide "how to manage water and its allocation, new development projects, pollution abatement and control restrictions, indeed all subjects dealing with water use" (Porto 1998, 177). She could use her knowledge of the seriousness and interlocking nature of water-related problems to try to find solutions, working with government officials and other water users. But does this actually happen? Before examining the possibility of poor rural women participating in this high-level decision-making body, this section explores women's participation in water management at the local and micro-regional level through the case of a rainwater-harvesting program called 'One Million Cisterns.'

Brazil has 12 percent of the world's freshwater. However, water resources are distributed unevenly in the country, and there are enormous regional discrepancies (Moraes and Perkins 2007). It was estimated that out of the one billion people of the world living in areas susceptible to desertification at the turn of the century, 25 million lived in the Brazilian semi-arid region (ASA 1999), located in the northeastern part of the country.

Two million families in this region live without adequate access to clean drinking water. Traditionally, because of its irregular rainfall, the semi-arid region falls victim to frequent droughts and floods. Rainwater harvesting (and dam development discussed later in this chapter) as well as adaptive measures like water reuse were attempts to even out water resource availability (Ribeiro Neto et al. 2014). As Biswas argues with respect to developing countries: "[T]o smooth out these wide inter and intra-annual fluctuations in rainfall . . . dams (small, medium and large), groundwater recharge and storage, as well as rainwater harvesting [must be considered] . . . to resolve . . . water problems" (2012, 11).

Development programs in this semi-arid region in Brazil are primarily focused on response to natural hazards and emergencies, aimed at fighting drought. In 1999, NGOs and other civil society organizations came together as an umbrella NGO called *Articulação no Semi-Árido*, or ASA. The newly formed network proposed to revolutionize the approach to development in the area by embracing, rather than fighting, its natural environment. Its founding document, the Declaration of the Semi-Arid Region,[6] proposed the "coexistence" of humans and nature within the semi-arid region as a way to prevent desertification (ASA 1999). Criticizing existing policies as inadequate to sustain the region's people and environment, the Declaration was built on the premise that there was a need for conservation and sustainable use of the semi-arid region's natural resources, along with a reduction in unequal access to land, water, and other means of production. The document proposed six main strategic actions,[7] among which the inclusion of women and youth in the development process took prominence.

The document recognized that: women account for 40 percent of the labor force in rural areas; they work, on average, 18 hours per day; more than half of all girls start working at the age of ten; and the majority of women are responsible for the water used within households.

The first program[8] developed by ASA under these principles was called 'Education and Mobilization Program for Coexistence in the Semi-Arid Region: One Million Cisterns,'[9] or simply P1MC. The main goal of the program was to build cement plate cisterns for one million families—half of the families without adequate access to clean drinking water in the semi-arid region of Brazil. Since 2003, the program has received support from the federal government as part of its Zero Hunger Strategy.[10] By January 2014, more than 500,000 rainwater harvesting cisterns were built, benefitting more than two million people (ASA n.d.).

Women are the main beneficiaries of the program, since the purpose of the cistern is to provide water for the household and women are assumed responsible for this task (see Figure 5.1). The cisterns built under the P1MC have the capacity to store 16,000 liters of water collected during the rainy season. Rainfall on roofs is collected by gutters, which conduct the water to the cistern. The water stored during the rainy season can last up to eight months for drinking and household use for a family of four. Families who receive the cistern participate in training about the Brazilian semi-arid region, and in water management

Figure 5.1 Woman and cistern, Northeast Brazil

Photo credit: José Luiz Oliveira (Codevasf—São Francisco and Parnaíba River Development Company)

(including maintenance of the cisterns) given by local agents hired and trained by the program. Some agents are also members of local or municipal program commissions. The selection of families that will receive cisterns in each community is done by local program commissions, following these criteria: (a) households headed by women; (b) number of children from zero to six; (c) children and adolescents in school; (d) presence of adults aged 65 or above; (e) presence of members with mental and physical disabilities.

The benefits of having a cistern as a source of drinking water and water for household use are quite self-evident, but it is worth describing the specific benefits to rural women in the Brazilian semi-arid region. First, as landowners control many wells and ponds in this region, women must wait many hours to be allowed access to collect water. Furthermore, the majority of women are fetching water 2, 4, 5, or 10 kilometers away. As one woman describes:

> They [women] are the ones responsible for getting water for making food. And when the cistern arrives, that time that she used to fetch water, that she had to wake up at 5 or 4 a.m., she no longer needs to wake up that early. Or if she wakes up, she will have time to do other things. So this is a big difference with the arrival of the cistern.
>
> (Selma, Cetra, Fortaleza, July 2008)

Also, since water is a basic necessity for cooking, washing, and cleaning, the access to water provided by the cistern directly benefits women's work in the household. The cistern is also an opportunity for household members to gain access to improved quality drinking water. Some state ASA forums (such as in Ceará) are already entering into partnerships with health agents so that families can continue to receive support in relation to water quality.[11]

It is clear that the P1MC is aimed at increasing women's access to water, making them its main beneficiaries. Originally, however, the program did not foresee an extended participation of women as decision-makers, or cistern builders (a paid position in the program).

The idea of women building cisterns in the P1MC was first conceived by members of a feminist organization, the *Casa da Mulher do Nordeste*,[12] and then embraced by a small number of rural women from the micro-region of Pajeú. Together with other supporters participating in ASA meetings, members of *Casa* raised the issue of gender equity in the program's implementation. Through an analysis of the gendered division of labor in the P1MC, they pointed out that the program was in fact reinforcing existing gender roles that helped maintain women in poverty. They highlighted two aspects in particular. First, poor rural women were under-represented in political decisions about the program, such as in the micro-regional commissions who selected communities to receive the cisterns. Second, it was assumed that women did not want to build cisterns and therefore they were not offered the opportunity for training or jobs to construct cisterns (Moraes and Rocha 2013).

Members of *Casa* argued that decision-making positions in the program, especially the participation of members of ASA's NGOs in municipal commissions, were positions of power in the semi-arid region and continued to be occupied mostly by men. Even though the management of domestic water directly affected rural women, it was very difficult for them to access decision-making positions. Leadership positions in organizations were perceived as mostly masculine and attracted more males, and they were also positions of power that men did not want to give up.

It was also noted that women were not offered the opportunity to work for pay in the program. The P1MC offered several "volunteer" opportunities, or in-kind contributions, to be provided by the family, and women participated fully in those. However, the people selected to train as masons to build the cement plate cisterns were invariably men. They were paid around BR$160 (about US$100) for each cistern, which was a sizeable amount in this region. It takes on average one week to build a cistern. When the women from *Casa* raised the question at the ASA meetings why women were not participating in the training to build cisterns, the response was that women had no interest. Members of *Casa* then asked rural women if they were interested and soon they had a list of candidates eager to learn how to build the cement plate cisterns (Moraes and Rocha 2013).

The first training course for women was offered in the Pajeú region. Out of six registered participants, three finished the course and started building cisterns.[13] Curiously, after the first female cistern builders were trained, they were not hired to build cisterns in their own micro-regions. Instead, the women trained in Pajeú were called upon to teach other women and build cisterns in other states under the sponsorship of feminist NGOs.

The second group of female cistern builders was linked to the March 8 Feminist Centre (CF8) in Mossoró. The CF8 supports and encourages the organization of rural women's groups in their region (Mossoró – Apodi). In one of their meetings the women (already in leadership positions in their own communities) said that one of the most important issues they wanted to address was water. In that area, water was very costly, even for communities with access to wells, because the maintenance of pumps was expensive. In partnership with the CF8, the women's group demanded cisterns for their communities. Through their discussions, the group decided women would participate in all stages of the project, including building the cisterns. CF8 members already knew of the case of the women in Pajeú and, through *Casa da Mulher do Nordeste*, they contacted the instructor who came to teach them.

A similar process unfolded in the micro-region of Fortaleza, where *Casa Lilás*, another feminist organization, invited the woman instructor from Pajeú to train more women to build cisterns. In Fortaleza, the coordination of the micro-regional management was by Esplar,[14] an NGO that was not only sensitive to gender, but also had women in leadership positions.[15] This enabled the implementation of a parity system of representation, where women occupied 50 percent of the positions on micro-regional water commissions. The participation of women in local commissions, and later in municipal and micro-regional commissions, had a significant effect on the lives of women and their communities, opening several possibilities for growth, knowledge exchange, and even political power. At the same time, their involvement depended on whether the organizations acting as micro-regional management units of the program were aware of and embraced gender equity; their involvement also depended on the engagement of women leaders in ASA who fought for their inclusion.

Up to 2011, ASA had trained more than 6,000 cistern builders (ASA n.d.). Gender disaggregated data are not available, but it is estimated that between 5 and 10 percent of cistern builders trained by ASA are women (Moraes and Rocha 2013). Furthermore, in 2009, the first woman was elected as President of the Association of the One Million Cisterns program—Valquiria Alves Smith Lima. Despite limitations and institutional barriers,[16] poor rural women have been increasingly participating in the program as NGO members, leaders, educators, program committee members, and cistern builders.

For the first women cistern builders it was difficult to find work in their own regions. Even when women were selected to build cisterns, they reported having to face community distrust due to their gender. At first the work was said to be too difficult or too heavy for a woman. Paradoxically, once the cisterns built by women became available and they did not leak, building cisterns was then seen as too easy for men—since "even a woman" could do it.

Another challenge experienced by women cistern builders was leaving home for prolonged periods of time in order to build cisterns in other communities. Although poor rural women do go out to fetch water, to work in the fields, or to work in other informal jobs (such as coal gathering), those activities normally do not mean leaving domestic responsibilities for long periods of time. Extended family support could, and many times did, help women to overcome this barrier. In a number of cases, in order to leave home, the cistern builders shared the care of their children and other household responsibilities with other women, such as mothers, sisters, and friends. There were also cases in which women shared domestic responsibilities with their male partners, bringing about changes in gender relations within the household.[17] Women also left the house to participate in other local, regional, and national meetings organized regularly by the P1MC as part of their mobilization strategy. Since 2008, cistern builders have had their own regional meetings, which are now part of the overall program budget.[18]

Despite these challenges, women cistern builders reported being proud and having a greater sense of capability and self-worth. They played a major role in building the cisterns, which became important community assets. This work increased their status as productive members of their communities. Before becoming cistern builders, these women had no recognizable profession outside the home, and their training and experience opened new professional opportunities. Financial gains were significant as the payment for the construction of cisterns was much higher than what they could earn doing informal work in mining, agriculture, or domestic services. Their experience also attracted media attention beyond local newspapers and some women appeared on a feature program on the national media network TV Globo. They became recognized in their communities as cistern builders.

Another positive element related to women's participation in the cistern building program was their closer connection to feminist organizations. These organizations actively included women cistern builders into a larger network of services (including water), opportunities (including work), and knowledge (formal and informal). For example, CF8 encouraged the cistern builder Linda to participate in a continuing education program offered to rural residents by the State University of Rio Grande do Norte. NGO members also referred cistern builders to new projects, such as the 'One Earth and Two Waters' program—which builds larger cisterns for agricultural production.

Rural women's connection to feminist organizations facilitated their participation in other collective action (such as the *Marcha das Margaridas* and the World Women's March) and networks (such as the Network of Women Producers of Pajeú,[19] and the Xique Xique Network for Solidarity in Mossoró[20]). Those connections helped to position their struggle for water as a collective struggle for economic and political inclusion, food security, social justice, and human and women's rights. However, could any of these women participate in water management at a larger scale?

The next section will discuss the Brazilian Water Law that created a participatory, decentralized, and territorial watershed management system, which includes the formation of Watershed Management Committees composed of members of civil society, government, and other users. In theory, poor rural women from the semi-arid region could participate in those committees, but in practice, this has not been easy for them to do.

Women's participation in the National Water Resources Management System[21]

In 1997, a law setting out the Brazilian National Water Resources Policy was approved (ANA 2006; Keck and Abers 2004; Ministério do Meio Ambiente 2005). The policy is based on six principles:

(1) Water is a public good.
(2) Water is a finite resource and it has economic value.
(3) In case of shortage, human consumption and animal watering are top priorities.
(4) Management must include multiple water uses.
(5) The watershed is the territorial management unit.
(6) Water management should be decentralized and include the participation of government authorities, users, and civil society (citizens).

The 1997 Water Law also created the National Water Resources Management System, with the purpose of implementing national water policy as well as coordinating integrated water management, arbitrating conflicts, planning, regulating, controlling use, preserving and recovering water resources, and encouraging the charging of fees for water use. The National Water Resources Management System is composed of the following institutions:

(1) The National Council of Water Resources, a consultative and deliberative body representing water users, government, and civil society. The council, among other responsibilities, coordinates federal, state, and regional planning and arbitrates conflicts.
(2) State Councils of Water Resources, which have similar responsibilities at the state level.
(3) Watershed Committees composed of state government officials, representatives of the municipalities, and members of civil society.
(4) Government agencies in water-related areas.
(5) The National Water Agency (ANA), linked to the Ministry of Environment, which implements the national water policy, grants water use permits, supervises, and charges for the use of water on a national scale.
(6) Watershed Agencies, which act as executive secretaries of the Watershed Committees, offering technical, financial, and administrative support for the implementation of their decisions.

Hence, Brazil has a well-developed and fairly progressive water management system that in principle provides space for civil society participation in watershed management—as prescribed by sustainable development theory. Women are present at all levels of the Brazilian water management system to some degree. The National Council of Water Resources, for instance, has current Environment Minister Izabella Monica Vieira Teixeira as its president, and has a seat for the Special Secretary for Women's Policies, Maria do Carmo Godinho (Secretaria Especial de Políticas para as Mulheres). Table 5.1 illustrates the current composition of the National Council for Water Resources, which had approved the incorporation of gender issues into its Base Reference Document (WWF 2006).

Table 5.1 shows that in 2013 women represented 36.5 percent of the total number of members in the National Council of Water Resources. This is an increase in representation from 2005 (Moraes and Perkins 2007) when women made up only 8.8 percent of the Council members. Although there has been visible progress in women's participation in politics in Brazil,[22] water management committees are more specialized in focus than other government bodies, and thus even harder for poor rural women to enter.

Water in Brazil has historically been linked with hydroelectric dams and energy, and therefore has traditionally been a subject area for engineers—a profession in Brazil that continues to be dominated by men. But the new Water Law emphasizes water as a multidisciplinary issue. University of São Paulo professor Monica Porto, an expert on Brazilian water management, had previously anticipated that women's participation as professionals in the water sector was growing very fast and, therefore, gender was not the main issue to be discussed.

Table 5.1 Composition of the National Council of Water Resources, by position, gender, and category, 2013

	Titular members		*Alternate members*		*Total*
	Men	*Women*	*Men*	*Women*	
National Government	24	6	24	6	60
State	5	5	8	2	20
Users[a]	8	4	8	4	24
Civil society[b]	4	2	4	2	12
Total	41	17	44	14	116
Women					31
Men					85

Source: Adapted from CNRH (2013)

Notes:

a The category users include: irrigators; providers of water and sanitation services; generators and providers of electricity; members from the waterways sector; industries; and fisherman and tourism users.

b The category civil society includes: members of committees and inter-municipal associations; technical education and research associations; and non-governmental organizations.

In her view, the need for a sufficient number of trained and competent professionals was an issue that prevails over gender *per se* (Porto 2000, 91). Porto believed that the successful implementation of the Water Law in Brazil depended on the ability of the Watershed Committees to: (a) use a transdisciplinary approach, (b) raise awareness about water issues among the public and policy makers, (c) educate the population, (d) prepare communities to participate, and (e) build technical capacity (Porto 2000). According to Porto:

> The country's poor has been constantly disregarded with respect to political decisions related to water. The poor have no access to safe water, have high infant mortality rates, and a very weak and usually unheard voice in asking for change.
>
> (2000, 90)

A study from 2010 examining gender roles in the Watershed Committees suggested that women who participate in the water management system tend to have higher levels of education (Empinotti 2010). Women comprised 55 percent of students registered in colleges and universities in Brazil in 2012 (INEP 2012). Hence, higher education has been a contributing factor in women's access to decision-making in highly technical spaces—such as the Watershed Committees.[23] However, Empinotti (2010) argues that women often perform different roles than men. Women tend to be more involved in logistics, such as in organizing events and preparing written documents, while men are more present in representative forums and trading activities. Therefore,

Table 5.2 Average years of education in Brazil, by gender and race, 1999 and 2009

	White women	*Black women*	*White men*	*Black men*
1999	8	5.6	7.1	4.7
2009	9.7	7.8	8.8	6.8

Source: Adapted from IPEA (2011)

Watershed Committees tend to reproduce a social dynamic where women play a supporting role, and men are the main decision-makers, even in cases where women have higher education and higher income.

Examination of the intersection of race, class, and gender in Brazil indicates that women participating in higher levels of decision-making, such as the Watershed Committees, are more likely to be white (see Table 5.2). Due to the pervasive colonial history of slavery in the country, the majority of the poor are still of African descent, despite advances towards more social equality. Will this translate into greater rates of participation of rural, poor, and, most likely black, women in the National Water Resources Management System? My research on the recent struggles of a Watershed Committee from the semi-arid region, and ASA's participation in it, suggests that the process is very complex.

Women's participation and non-participation in the Lower São Francisco Watershed Committee

The São Francisco River watershed comprises more than 500 municipalities and six states, mostly in the northeast region of Brazil (see Figure 5.2). The river's source is in the mountains of Minas Gerais, from which it runs north passing through the states of Bahia, Goiás (including the Federal District), and Pernambuco. Then the river turns east and crosses the borders of the Alagoas and Sergipe states, then flows into the Atlantic Ocean.

The São Francisco River extends over 2,700 km and is a critical water source for the northeast region which receives a highly variable 200 to 700 mm of rainfall annually (Braga et al. 2012). It provides 70 percent of the water for the northeast region; it is host to eight electrical plants, the Sobradinho Dam (the largest in Latin America) and its cascading water reservoirs (Empinotti 2011). The number of watershed inhabitants is about 14.2 million people, or 7.5 percent of the total Brazilian population (ANA n.d.), and is the poorest in Brazil (Braga et al. 2012). The region is marked by several contrasts in terms of population density and wealth. It comprises large metropolitan regions, such as Belo Horizonte, as well as dispersed rural areas. Diversity also characterizes the human uses of the river, which includes agriculture and irrigation, energy generation, sanitation, fisheries and aquaculture, tourism and leisure, and navigation. Only 98 of its municipalities have sewage treatment (ANA n.d.) and water pollution is a severe problem, in addition to siltation, deforestation, and erosion. The Old Chico (as the São Francisco River is called by locals) has

Figure 5.2 Location of the São Francisco River (Rio) watershed within Brazil and map of Brazil

Source: Gary Christopherson, University of Arizona, School of Geography and Development, Spatial Database Development and Cartographic Design

retained only 4 percent of its original riparian forest (Diario de Pernambuco 2012). Projections for northeast Brazil are for increased dryness (Krol et al. 2011; Seneviratne et al. 2012) due mainly to climate change, thus water demand in the São Francisco River basin is likely to increase while flows in the river are likely to be reduced.[24]

The São Francisco Watershed Committee was officially approved in 2001, with its first members elected in 2002. Members of ASA from the state of Alagoas became interested in participating and after mobilization, elected one ASA member as an alternate representative on the committee in the civil society sector (Empinotti 2007). ASA's members found that the committee could be another tool for disseminating their model of coexistence with the semi-arid region. They had already experienced success in this region, getting improved support from the national government for the P1MC—that had started a pilot project in 2000.

One of the main issues discussed in this Watershed Committee was the proposal for the diversion of the São Francisco River—a mega project that generated heated debate in the region (COPPE 2005). For ASA members, the São Francisco River Diversion Proposal was another symbol of the modernization model of development that was going to benefit the irrigation interests of large-scale farms and agribusiness. The diversion of the São Francisco River, as with other previous large-scale projects, was not going to resolve the water insecurity problem of a vast number of dispersed small farmers in the region (Empinotti 2007, 181). Soon ASA members realized that the government was going to implement this project anyway, and became aware of the limited role of the São Francisco Watershed Committee in decision-making. The diversion project was approved at an even higher scale through the National Council of Water Resources, against the recommendations of this regional Watershed Committee.

When new elections for the Committee took place in 2005, ASA members decided not to participate. Their struggle for the coexistence with the semi-arid region would benefit more from a direct relationship with the national government (through programs such as One Million Cisterns) than through the formal institutional water management committee. According to Empinotti (2007) this was a strategic move—where non-participation did not mean resignation, but a choice to participate in water management through the direct partnerships with the national government.

This case illustrates the complexity of participation in water management in Brazil, and that the increased participation across different levels is not linear—in fact, it is composed of many advances and setbacks for women. In 2013, the Lower São Francisco Watershed Committee had 105 members, of which 81 are men and 24 are women. From the total number of women, 14 are titular members and 9 are alternates (CBHSF n.d.). Women are still a minority.

Discussion and conclusions

This chapter began by identifying a paradox concerning women's participation in water management: while women in poor rural regions of the developing world are often responsible for water management at the household level, women are very under-represented in water management decision-making processes. An intersectionality approach framed this paradox in the case of access to water—it is mostly poor rural (and black) women who struggle to access and manage clean water, yet at the same time these women are excluded from decision-making on water management at higher levels. This reveals a context in which not only gender inequalities matter, but also inequalities between different groups of women.

The examination of the continuing expansion of women's roles in the programs coordinated by the ASA—such as the One Million Cisterns program—shows an optimist scenario. Despite the long *macho* tradition in the northeast region of Brazil, the ideology of coexistence with the semi-arid region and its resources (environment and people) seems to converge with the increased value of women's rights in the region. Feminist organizations have been gaining space in the P1MC and are acting as effective agents of gendered social change (Goulet 1989, 2005). Feminist organizations have supported poor rural women's mobilization and organization at the local level, and helped women to take advantage of new spaces of participation in the program as commission members and cistern builders. In addition, feminist organizations on the ground have been working as network hubs that share with poor rural women information about other opportunities—such as training, work, and even education. Despite the existing mainstream gender and racial inequalities in water management in the Brazilian semi-arid region, the growing participation of poor rural women at the community and regional levels is now visible and promises to continue. The synergy of recent policies and programs increasingly accessible to poor rural women also reinforces this trend.[25]

The second case examined in this chapter referred to the overall participation of women in institutionalized and official spaces occupied by organized civil society with respect to watershed decision-making in Watershed Committees and the National Water Resources Council. It was suggested that access to education is a path towards greater participation of women: the more education a women has, the greater her chances for participating in these regional and national spaces of decision-making about water. However, Brazil inherited from its colonial past huge social inequalities, which are still evident when comparing years of education between blacks and whites in the country, leaving black women disadvantaged compared to whites and men. Moreover, despite some progress, gender inequality remains significant at higher levels of water management decision-making, with more men occupying visible positions of power, and more women playing supporting roles.

Up to this point, the line of thought regarding the progression of women's participation in water management seemed to be linear: first, by recognizing that poor rural women do manage water at the household level; second, by acknowledging that in participatory programs such as P1MC women now have greater chances to participate in decision-making about water at the community level. This is especially important in light of projections of increased dryness in northeast Brazil due to climate change, as drought and less water availability will exacerbate water-related challenges. At what seemed to be the next logical level, the Watershed Committees, a closer look at a specific committee revealed more complexity, suggesting that higher levels of power in water management are still a big challenge for the poor in general, and women in particular. Moreover, these cases illustrate how power is gendered in every society, and, as a result, the management of water is as well.

Notes

1 For instance, different needs in terms of privacy and hygiene, physical safety (when and where to get water, relieve oneself), preferred crops, as well as resources, responsibilities, networks.
2 Migration is a very common livelihood strategy in drought risk areas, such as the Brazilian semi-arid region.
3 First, they use irrigation not only for crops, but also for gardens and animals. Second, they use different types of water (quality and quantity) for different purposes, as already suggested by van Wijk-Sijbesma (1998). Third, because of their time commitments for domestic roles, women have less flexibility than men in terms of time.
4 This is true, for instance, with the Watershed Committees created in Brazil, which are open to civil society participation. Women participate less than men, and when they do participate, it is often in subordinate positions (Moraes and Perkins 2007). Although there have been improvements, the participation of low-income women, and especially non-white women, is still a big challenge.
5 Excerpts of this section were previously published in Moraes and Rocha (2013).
6 The Declaration of the Semi-Arid Region was the founding document of ASA in 1999 and is adopted by all its members. It is available at www.asabrasil.org.br/Portal/Informacoes.asp?COD_MENU=104. Accessed January 12, 2006.
7 (1) To coexist with drought; (2) to guide investments towards sustainability; (3) to strengthen civil society; (4) to include women and youth in the development process; (5) to maintain, rehabilitate, and manage natural resources; and (6) to finance the program for coexistence in the semi-arid region (ASA 1999).
8 ASA later developed a program called 'One Earth and Two Waters' (*Uma Terra e Duas Águas*) that aims at constructing larger cisterns to collect water for production. It also started a pilot project constructing school cisterns.
9 In Portuguese: *Programa de Formação e Mobilização Social para a Convivência com o Semi-Árido: um Milhão de Cisternas Rurais*. See www.asabrasil.org.br/Portal/Informacoes.asp?COD_MENU=1150. Accessed January 14, 2006.
10 For an overview of the Zero Hunger Strategy see Rocha (2009).
11 For a discussion of water quality issues related to cistern use in the region refer to Gould (1999) and Freitas et al. (2011).
12 *Casa da Mulher do Nordeste* is a feminist NGO founded in Recife, Pernambuco in the 1980s. *Casa*'s main goal is "to contribute to the productive and political action of poor women in the Northeast, towards gender equity in a perspective of sustainable human development" (www.casadamulherdonordeste.org.br/index.html). *Casa* has two main programs: 'Gender and Economy' and 'Women in Rural Life.' It maintains two offices, one in Recife and another in Afogados da Ingazeira in the region of Pajeú. *Casa* has several national and international partners and is connected with a number of networks. *Casa* is also a member of ASA and participates in ASA's national meetings, as well as in meetings of the State of Pernambuco and of the micro-region of Pajeú. It was through this participation at ASA meetings that members of *Casa*, in partnership with other feminists, began to raise some questions about gender equity in the P1MC.
13 Some level of attrition was typical for courses for men as well.
14 For more information about the NGO Esplar see www.esplar.org.br/quem/oque.htm. Accessed March 22, 2004.
15 The position of coordinator of the micro-regional unit of the P1MC in Fortaleza was consecutively occupied by women who had leadership positions in the State ASA itself (Forum Ceará).
16 See Soares (2009).
17 To illustrate: the minimum salary in Brazil in 2004 was around BR$250, which was common in the region. A woman building four cisterns a month at BR$150 each

could make more than double this amount. One can imagine how this could have affected gender relations.

18 See the budget at www.diariosoficiais.com/home/ler_diario_index?data=28%2F02%2F2011&diario=2&pdf=dou_s3_2011_02_28_arq_1_pg_00160.html&sesSão=3. Also, a couple of examples are the meetings of cistern builders of Ceará and Rio Grande do Norte (2009) http://solidu.wordpress.com/2009/04/20/128/; Minas Gerais, Sergipe, and Bahia (2010) www.cavjequi.org/index.php?pagina=8&id=84. Accessed March 22, 2004.

19 See *Rede de Mulheres Produtoras do Pajeú*: http://cirandas.net/rede-de-mulheres-produtoras-do-Pajeú. There are also videos about them available at www.youtube.com/watch?v=s_X1PctSmoY. Accessed March 22, 2004.

20 *Rede de Comercialização Solidária Xique Xique* www.redexiquexique.blogspot.com/. Accessed March 23, 2004.

21 Excerpts of this section were previously published in Moraes and Perkins (2007).

22 This culminated with the election and re-election of Dilma Roussef as the first female president.

23 Data about the numbers of students registered by color are not as easy to interpret. Recent statistics for 2012 (INEP 2012) report that approximately 23 percent are white, 3 percent black, 0.9 percent yellow, 11 percent parda (mixed race), 0.1 percent Indian, 35 percent without information, and 27 percent was not declared. The fact that 62 percent have no information invalidates the results. However, due to the colonial history of the country and long-lasting slavery, it is possible to assume that the majority of graduates from higher education are white, despite recent changes in educational policies.

24 For a map that shows the areas occupied by northeast Brazil, the semi-arid area within the northeast, and the São Francisco basin within the semi-arid area, see Braga et al. (2012, 155).

25 See for example women's increasing participation in Social Policy programs such as PRONAF (Funds for Family Farming Program), PAA (Food Acquisition Program), PNAE (National School Meals Program), *Conselho Tutelar* (Children's Rights Council), etc.

References

ANA (Agência Nacional das Águas). 2006. The evolution of water resources management in Brazil. Brasilia: Agência Nacional das Águas. www.ana.gov.br/ingles/Portais/06-General_Aspects.html#3. Accessed April 15, 2014.

——. n.d. Região hidrográfica do São Francisco: Águas que contribuem para o desenvolvimento de 521 municípios. Brasilia: Agência Nacional das Águas. www2.ana.gov.br/Paginas/portais/bacias/SãoFrancisco.aspx. Accessed April 15, 2014.

ASA (*Articulação no Semi-Árido*). 1999. Declaração do semi-árido Brasileiro. Recife, Brazil. www.asabrasil.org.br/Portal/Informacoes.asp?COD_MENU=104. Accessed April 14, 2014.

——. n.d. P1MC: Resultados. Recife, Brazil. www.asabrasil.org.br/. Accessed April 14, 2014.

Biswas, A. 2012. Impacts of large dams: Issues, opportunities and constraints. In *Impacts of large dams: A global assessment*, eds C. Tortajada, D. Altinbilek, and A.K. Biswas, 1–18. Berlin and Heidelberg: Springer.

Braga, B.P., J.G.C. Gondim Filho, M.R. von Borstel Sugai, S.V. da Costa, and V. Rodrigues. 2012. Impacts of Sobradinho Dam, Brazil. In *Impacts of large dams: A global assessment*, eds C. Tortajada, D. Altinbilek, and A.K. Biswas, 153–170. Berlin and Heidelberg: Springer.

Buechler, S. 2005. Women at the helm of irrigated agriculture in Mexico: The other side of the male migration. In *Opposing currents: The politics of gender and water in Latin America*, eds V. Bennett, S. Davila-Poblete, and M. Nieves Rico, 170–189. Pittsburgh, PA: University of Pittsburgh Press.

CBHSF (Comitê da Bacia Hidrográfica do Rio São Francisco). n.d. *A bacia: principais características, conflitos atuais e cenário desejável*. http://cbhSãofrancisco.org.br/. Accessed April 12, 2014.

CNRH (Conselho Nacional de Recursos Hidricos). 2013. *Composicao do CNRH*. www.cnrh.gov.br/index.php?option=com_content&view=article&id=99:composicao-cnrh&catid=1:o-conselho-nacional-de-recursos-hidricos-cnrh. Accessed April 12, 2014.

Collins, P.H. 1998. It's all in the family: Intersections of gender, race, and nation. *Hypatia* 13: 62–82.

COPPE (Instituto Alberto Luiz Coimbra de Pós-Graduação e Pesquisa de Engenharia). 2005. Transposição do Rio São Francisco: Prós e contras. Seção notícias. October 10. www.planeta.coppe.ufrj.br/artigo.php?artigo=636. Accessed April 25, 2014.

Crenshaw, K. 1991. Mapping the margins: Intersectionality, identity politics, and violence against women of color. *Stanford Law Review* 43: 1241–1299.

Crow, B. 2001. Water, gender and material inequalities in the global south. Working paper series from the Center for Global, International and Regional Studies and the Department of Sociology from the University of California Santa Cruz. http://repositories.cdlib.org/cgirs/CGIRS-2001-5/. Accessed September 4, 2013.

Diario de Pernambuco. 2012. *Pesquisadores anunciam a extinção inexorável do Rio São Francisco*. September 25. www.diariodepernambuco.com.br/app/noticia/vida-urbana/2012/09/25/interna_vidaurbana,398575/pesquisadores-anunciam-a-extincao-inexoravel-do-rio-São-francisco.shtml. Accessed September 8, 2013.

Elmhirst, R. 2011. Introducing new feminist political ecologies. *Geoforum* 42: 129–132.

Empinotti, V. 2007. Re-framing participation: The political ecology of water management in the lower São Francisco River Basin—Brazil. PhD diss, University of Colorado at Boulder.

——. 2010. Gênero, recursos hídricos e tomada de decisão: O papel das mulheres nos organismos de bacia Brasileiros. In *Água e política: Atores, instituições e poder nos organismos colegiados de bacia hidrográfica no Brasil*, ed. R.N. Abers, 161–190. São Paulo: Annablume.

——. 2011. E se eu não quiser participar? O caso da não participação nas eleições do comitê de bacia do Rio São Francisco. *Ambiente e Sociedade* 14: 195–211.

Freitas, D., M. Morais, J. Cabral, J. Rosado, G. Melo, H. Silva, A. Pedro, and V. Selva. 2011. Water quality of cisterns and underground dams in semiarid regions: Case study of Afogados da Ingazeira, Pernambuco–Northeast Brazil. IWA Conference Recife, Brazil. http://dspace.uevora.pt/rdpc/bitstream/10174/3739/1/O-059.pdf. Accessed August 13, 2014.

Gould, J.E. 1999. Is rainwater safe to drink? A review of recent findings. In 9th International Rainwater Catchment Systems Conference.

Goulet, D. 1989. Participation in development: New avenues. *World Development* 17: 165–178.

——. 2005. Global governance, dam conflicts, and participation. *Human Rights Quarterly* 27: 881–907.

Haraway, D. 1991. *Simians, cyborgs, and women: The reinvention of nature*. New York: Routledge.

Harris, L. and W. Gantt. 2007. Gender and the shifting water governance: Differential effects of privatization, commodification and, democratization. Tenure Brief, no. 6. Madison: Land Tenure Center, University of Wisconsin-Madison. http://minds.wisconsin.edu/bitstream/handle/1793/22157/%21ltcbrief6-water%20governance.pdf?sequence=1. Accessed September 9, 2013.

House, S. 2003. Easier to say, harder to do – gender, equity and water. Paper submitted to the Alternative Water Forum, May 1–2, Bradford Centre for International Development, University of Bradford. www.splash.bradford.ac.uk/files/PDF%20Alternative%20Water%20Forum%20House.pdf. Accessed July 25, 2012.

INEP (Instituto Nacional Estudos e Pesquisas Educacionais Anisio Teixeira). 2012. Table 5.4 – Número de matrículas em cursos de graduação presenciais, por organização acadêmica e sexo. In *Sinopses estatísticas da educação superior – graduação*. Brasilia: INEP. http://portal.inep.gov.br/superior-censosuperior-sinopse. Accessed September 1, 2013.

IPEA (Instituto de Pesquisa Econômica Aplicada). 2011. Retrato das desigualdades de gênero e raça. In participation with: ONU Mulheres, Secretaria de Políticas para as Mulheres (SPM), Secretaria de Políticas de Promoção da Igualdade Racial (Seppir). 4ª ed. Brasília. 39 p.

Keck, M. and R. Abers. 2004. Running water: Participatory management in Brazil. *NACLA Report on the Americas* 38 (July/August): 29–34.

Kevany, K., M. Siebel, K. Hyde, D. Nazer, and D. Huisingh. 2013. Water, women, waste, wisdom and wealth – harvesting the confluences and opportunities. *Journal of Cleaner Production* 60: 4–10.

Krol, M.S., M.J. de Vries, P.R. van Oel, and J.C. de Araújo. 2011. Sustainability of small reservoirs and large scale water availability under current conditions and climate change. *Water Resource Management* 25: 3017–3026.

McCall, L. 2005. The complexity of intersectionality. *Signs: Journal of Women and Culture and Society* 30(3): 1272–1300.

Ministério do Meio Ambiente. 2005. Documento de introdução: Plano nacional de recursos hídricos – iniciado um processo de debate nacional. Brasília: Ministério do Meio Ambiente. http://pnrh.cnrh-srh.gov.br/. Accessed June 17, 2008.

Moraes, A. and E. Perkins. 2007. Women, equity and participatory water management in Brazil. *International Feminist Journal of Politics* 9: 485–493.

Moraes, A. and C. Rocha. 2013. Gendered waters: The participation of women in the 'One Million Cisterns' rainwater harvesting program in the Brazilian Semi-arid region. *Journal of Cleaner Production* 60: 163–169.

Parpart, J. 2009. Fine words, failed policies: Gender mainstreaming in an insecure and unequal world. In *Development in an insecure and gendered world*, ed. J. Leckie, 51–70. London: Ashgate Publishers.

Porto, M. 1998. The Brazilian water law: A new level of participation and decision making. *Water Resources Development* 14: 175–182.

——. 2000. Women and water resources research in Brazil. In *Women and water management: The Latin American experience*, ed. C. Tortajada, 83–92. Oxford and New York: Oxford University Press.

Rap, E., P. Wester, and L.N. Perez-Prado. 1999. The articulation of irrigation reforms and the reconstitution of the hydraulic bureaucracy in Mexico. Paper presented at the International Researchers' Conference 'The Long Road to Commitment: A Socio-Political Perspective on the Process of Irrigation Reform?' December 11–14, 1999. Hyderabad, Andhra Pradesh, India.

Ribeiro Neto, A., C.A. Scott, E.A. Lima, S.M.G.L. Montenegro, and J.A. Cirilo. 2014. Infrastructure sufficiency in meeting water demand under climate-induced socio-hydrological transition in the urbanizing Capibaribe River Basin – Brazil. *Hydrology and Earth System Sciences (HESS)*. doi:10.5194/hessd-11-2795-2014.

Rocha, C. 2009. Developments in national policies for food and nutrition security in Brazil. *Development Policy Review* 27: 51–66.

Rocheleau, D. 1995. Maps, numbers, text and context: Mixing methods in feminist political ecology. *Professional Geographer* 47: 458–467.

Rocheleau, D., B. Thomas-Slayter, and E. Wangari, eds. 1996. *Feminist political ecology: Global issues and local experiences*. London and New York: Routledge.

Roschelle, A., M. Toro-Morn, and F. Elisa. 2004. Towards a feminist methodological approach to studying the intersection of race, class and gender in Cuba. Paper presented at the annual meeting of the American Sociological Association, San Francisco, August 14–17.

Seager, J., ed. 2008. Expert group meeting on gender-disaggregated data on water and sanitation. Knowledge No. 1. UNW-DPC Publication Series & UN-Water Decade Programme on Capacity Development. Bonn: United Nations University. www.ircwash.org/sites/default/files/Seager-2009-Gender.pdf. Accessed September 1, 2013.

Seneviratne, S.I., N. Nicholls, D. Easterling, C.M. Goodess, S. Kanae, J. Kossin, and X. Zhang. 2012. Changes in climate extremes and their impacts on the natural physical environment. In *Managing the risks of extreme events and disasters to advance climate change adaptation: Special report of the intergovernmental panel on climate change*, ed. C.B. Field, 109–230. New York: Cambridge University Press.

Soares, D.N. 2009. *Gênero e agua – Desenhos do norte, alternativas do sul: Analise da experiência do semi-árido Brasileiro na construção do desenvolvimento democrático*. PhD diss, Universidade de Brasília. http://repositorio.bce.unb.br/bitstream/10482/8055/1/2009_DanielaNogueiraSoares.pdf. Accessed September 1, 2013.

Truelove, Y. 2011. (Re-)conceptualizing water inequality in Delhi, India through a feminist political ecology framework. *Geoforum* 42(2): 143–152.

U.N. (United Nations). 2008. The millennium development goals report. New York: United Nations. www.un.org/millenniumgoals/2008highlevel/pdf/newsroom/mdg%20reports/MDG_Report_2008_ENGLISH.pdf. Accessed September 1, 2013.

U.N. Women (United Nations Entity for Gender Equality and the Empowerment of Women). 2011. Gender issues: Women, poverty & economics. www.unifem.org/gender_issues/women_poverty_economics/. Accessed September 1, 2013.

van Wijk-Sijbesma, C. 1998. *Gender in water resources management, water supply and sanitation: Roles and realities revisited.* The Hague: IRC-International Water and Sanitation Centre.

Wallace, T. and A. Coles. 2005. Water, gender and development: An introduction. In *Gender, water and development*, eds A. Coles and T. Wallace, 1–20. Oxford: Berg Publishers.

WWF. 2006. *Brazil approves national water plan*. News & stories, February 8. http://wwf.panda.org/wwf_news/?58760/brazil-approves-national-water-plan. Accessed June 17, 2008.

Zwarteveen, M. and V. Bennett. 2005. The connection between gender and water management. In *Opposing currents: The politics of water and gender in Latin America*, eds V. Bennett, S. Davila-Poblete, and M. Nieves Rico, 13–29. Pittsburgh, PA: University of Pittsburgh Press.

Part II

Women and innovative adaptation to global environmental change

6 Climate-water challenges and gendered adaptation strategies in Rayón, a riparian community in Sonora, Mexico

Stephanie Buechler

This chapter explores the creative tactics employed by dairy producers and women cheese makers to adapt to climate change and growing water scarcity in Rayón, a small community in Sonora, Mexico. The agricultural production and processing activities undertaken in this region are set in the context of temperature variability, surface and groundwater decline, and land degradation and conversion.[1] Riparian landscapes in northern Sonora offer critical livelihood and ecosystem services; livestock production on uncultivated land in these arid oases contributes to food security and income. This riparian context and the gender focus of the research combine to fill a gap in the literature on these topics, revealing how women's adaptation strategies in urban and rural spaces are linked via hydrological, ecological, political-economic, and gendered social networks.

Women continue to be made most vulnerable to environmental change by structural inequalities: their exclusion from household and policy-related decisions concerning resource use amplifies climate-related risks. The related challenges faced by this community are broadly representative of global trends in both arid and semi-arid lands. By 2030, accessible fresh water runoff is unlikely to have increased more than 10 percent, while the Earth's population is expected to have grown by 33 percent (Jackson et al. 2001). In addition, it is projected that 50 percent more food and 30 percent more water will be needed by 2030 in order to feed the Earth's population (Hanjra and Qureshi 2010). Mexico's food security is particularly at risk as approximately 17.6 percent of the population is currently food insecure and climate change will likely make food security deteriorate further (Lankao and Smith 2014). Despite this, women's critical roles in food provisioning continue to be inadequately accounted for in policy responses at local, national, and international levels (Momsen 2009).

Women's capacities to produce and process agricultural commodities to feed themselves and others depend on their ability to access and utilize natural resources. As Lorena Aguilar stresses:

> [the] disproportionate burden of climate change on women can be countered by their empowerment and recognition. Women have important primary roles as managers of forests, land, water, and other natural resources in many

> communities. This position makes them powerful agents of change in formulating responses to climate change.
>
> (2013, 150)

Women's empowerment and their recognition as important actors in agriculture can be achieved through greater appreciation for gendered knowledge, as feminist political ecology highlights. The social (i.e. gender, age, ethnicity, race) and physical locations of these actors have been examined to gain a better understanding of how social standings influence the ability to use gendered environmental knowledge to gain political voice and decision-making power within households and communities (Elmhirst 2011; Nightingale 2011). Although existing literature primarily considers women as the keepers of seeds (Cabrera et al. 2001; Carney 2001; Oakley and Momsen 2007; Parasecoli 2010; Sachs et al. 1997) or as field laborers (Barndt 2002; Gunewardena 2010; Momsen 2009), their involvement in post-harvest processing stages remains understudied and is therefore excluded from policy discussions concerning climate impacts on women and agriculture (Buechler 2009, 2012).

In order to better understand social systems in the region, the natural context must be provided. The San Miguel River area of northern Mexico (the context of this study) has uneven rainfall across states influencing surface water flows. Northern Mexico makes up 30 percent of the Mexican territory but generates only 4 percent of surface water flow in intermittently wet creeks and rivers.[2] Water for agricultural irrigation comes from rainfall-dependent springs, wells, canals, or river tributaries. Precipitation and temperatures in Sonora are highly variable. Half of the region's precipitation falls in July and August during what is termed the North American monsoon. Monsoon rains are usually strong, localized, and of short duration. In contrast, winter rains enable groundwater recharge, heavier river flows, and increased mountain snow pack, which feeds rivers in spring (Vivoni et al. 2007b). The average annual temperature in the Sonora River basin (the study area, see Figure 6.1) is 75.9°F, ranging from 32°F or below in the winter to more than 113°F in the summer (CEA Sonora 2008). Cook and Seager (2013) predict later-onset monsoons and the Intergovernmental Panel on Climate Change (IPCC) predicts temperature increases of up to 3–5°C by 2100, with possible decreases of 5–8 percent in rainfall for the northwestern Mexico-U.S. border region (Lankao and Smith 2014; Wilder et al. 2010).

Groundwater over-draft (over pumping) and surface water overuse will be exacerbated by these climatic changes as well as by water governance limitations. The three Sonoran River basin councils (*consejos de cuenca*) and the COTAS (*Comité Técnico de Aguas Subterráneas* or Technical Committee on Groundwater),[3] an auxiliary arm of the river basin council and the CONAGUA (*Comisión Nacional del Agua*),[4] which focuses on agricultural groundwater users in the Rayón region's San Miguel River basin, have thus far had their efforts stymied to control groundwater use due to insufficient funding, too few staff, and CONAGUA's centralized decision-making (Wilder 2008).[5]

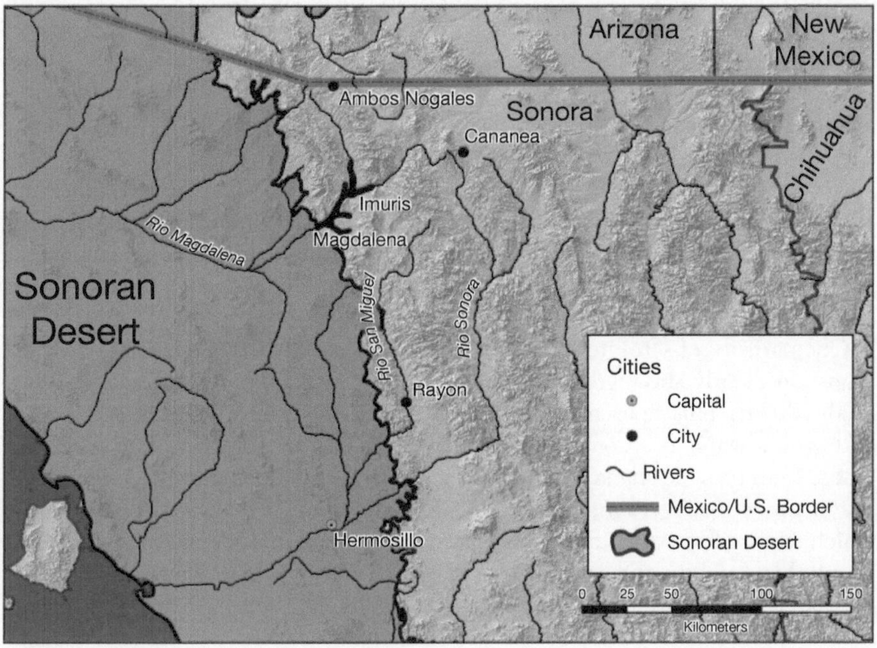

Figure 6.1 Map of Sonora showing location of Rayón along the San Miguel River

Source: Database development and cartographic design by Gary Christopherson, University of Arizona for use by the author, Buechler

Rayón is located along the San Miguel River (Rio) (see Figure 6.1), a large tributary to one of Sonora state's seven main rivers, the Sonora River (Padilla Calderón 2012). The San Miguel watershed is approximately 4,000 km^2 with a population of 5,092 people in 2010 (Moreno Vasquez and Navarro Navarro 2012). Rivers in the region flow mainly after winter rains, during the monsoon season and in October if Pacific hurricanes produce rain. Mean annual precipitation in the San Miguel River sub-watershed is 400–500 mm (approximately 16–17 inches). The watershed has been declared as zero or in deficit in terms of the balance between volume of available surface water and volume allocated to downstream use (CONAGUA 2009). The San Miguel and other rivers could be described as disappearing rivers due to increased human appropriation of their flows, declining rainfall, and greater evaporation of existing flows.

Part of the San Miguel River's flow is diverted to the Abelardo L. Rodriguez Dam, which supplies water to Hermosillo, the capital city with 715,061 inhabitants.[6] With rapidly growing cities like this one, rural areas such as Rayón are affected by larger rural-urban river water transfers. Such transfers are common globally: rural areas lose water to larger population centers, often with gendered effects.

The vegetation of the San Miguel watershed is a subtropical scrubland ecosystem known as the Sinaloan Thornscrub including organ pipe cactus, mesquite, tree ocotillo, palo verde, palo blanco, and whiteball acacia (Vivoni et al. 2007b). Cattle are kept in fenced areas within the scrubland, referred to locally as the *monte*, and graze on the leaves of the vegetation. This scrub area is expanding on both sides of the Mexico-U.S. border as temperatures increase and rainfall declines. This land has experienced great change with forested areas cleared, followed by grasslands predominating, and then a conversion to scrubland. Due to the vegetation's limited temperature range and water requirements, the main factor impacting it today is climate change (House-Peters and Chan 2014).

Comparisons of satellite images of northwestern Mexico taken pre- and post-monsoon clearly show greening of vegetation, especially along rivers (Vivoni et al. 2007a). This riparian vegetation is changing because of increased surface and groundwater use combined with less rainfall: cottonwood tree seedlings (these trees provide shade for cattle), for example, are unable to survive because they cannot reach groundwater. Mesquite, tamarisk, and other woody plants, which require less water, take their place (Hough et al. n.d.), making the area less hospitable for humans and their livestock. Woody plant encroachment on grasslands and riparian vegetation has been documented in other river basins such as the Edwards Plateau region of Texas and has been termed a process of desertification (Wilcox and Huang 2010).

Agriculture in Mexico and Sonora

Agriculture is an important sector of the Mexican economy. Fifteen percent of its 50.4 million economically active population (7.7 million) worked in agriculture in 2010, 12 percent of whom were women. This figure is likely higher as women's participation in family agricultural enterprises is often under-counted in agricultural census data (IUCN 2013).[7] Sonora is among the top six agricultural-producing states in Mexico and second in area of irrigated land: in 2010 it had 658,509 hectares under irrigation (11 percent of Mexico's total).[8] This land is irrigated with water from rivers, springs, and groundwater, however, the climate and water dynamics described above are nearing threshold levels beyond which irrigated agriculture may not be able to be practiced. This situation is similar to those documented on riparian areas in the west and southwest U.S. (Garfin et al. 2013; MacDonald 2010; Vásquez-León et al. 2003; Welsh et al. 2013; Wilder et al. 2010).

Dairy production and processing in Rayón

For this study, which covers the period January 2012 to January 2014, 20 cheese makers plus ten ranchers were interviewed through in-depth interviews; seven follow-up interviews were also conducted of the cheese makers to get a better sense of the environmental, economic, and political changes in their enterprises over time. These methods also reveal dynamism in gendered strategies of household members to address climate- and water-related challenges. Informal discussions were also held with governmental and non-governmental agency staff and private well-drilling company staff in Rayón and for the San Miguel River watershed. A research project stakeholder workshop on climate change and ranching was held in January 2014 and responses of ranchers and government officials were recorded. The stakeholder meeting allowed for observation of discussions between community members, particularly ranchers, and government agency staff and political figures.

The municipality of Rayón has five main villages: Rayón, La Galera, Tres Alamos, La Paz, and Cerro de Oro with a total population in 2010 of 1,599.[9] The town of Rayón had a population of 1,382 with agricultural fields and small ranch houses located in these other localities. The San Miguel River is the main surface water stream crossing the municipality. Water for irrigation comes directly from this river that is channeled into canals and flows during and right after summer and winter rains, and also comes from groundwater sources. Lower river flows are often mentioned in the interviews as with one dairy producer who shook his head and said: "The river is usually dry . . . since 1994 there have not been flood-related rises in the river water levels; many of the fields along the river have been abandoned." A woman rancher and cheese maker who was asked during a project stakeholder meeting if she thought that flooding was a risk, replied emphatically "I do not think that is a problem because it has not rained much in fourteen years!"

Groundwater is pumped from wells all along the river that are often operated as group wells with 5–15 farmer members sharing the water and the well maintenance and pumping costs. The water is often piped several kilometers from the well to the fields it irrigates. Many ranchers have dug wells that are only deep enough to supply water for their cattle, not for the irrigation of their crops; these are usually 60–80 feet deep. Water for domestic use also comes from wells. In spring 2013, the main well was repositioned, meaning that a new well was drilled that is 130 feet deeper than the dug well that it replaced (well-drilling team manager, interview, March 2013).

In 2004, agriculture in Rayón was developed on 1,993 hectares; 1,720 of these are irrigated, and 273 are rain fed (Plan Municipal de Desarrollo de Rayón 2004). Most agriculture is devoted to the production of forage crops that include rye grass, oats, alfalfa, and barley that are irrigated with groundwater and/or river water channeled in canals or crops like buffel grass (*Cenchrus ciliaris* or African foxtail grass) on rain fed land. Cattle-rearing for meat and dairy production is the main economic activity. In 2011, there were almost twice as many cattle kept for dairy production as for meat production; some cattle are used for milking and then sold to the local or regional butcher shops or meat packing plants when they reach the age of 5–10 years (INEGI 2011). Forage production and livestock-rearing are developed on approximately 108,200 hectares; 74 percent is owned by small private landholders, 18 percent are communal lands, and 8 percent is *ejido* land.[10]

Climate change is impacting this dairy production in myriad ways. Temperature extremes are more common now. As one rancher said, "it can change drastically in temperature in a 24 hour period" (stakeholder workshop, interview, January 2014). The livestock feed is impacted by winter freezes, high summer temperatures, and by drought. For half the year or longer, the cattle are allowed to graze in the *monte* (scrubland desert) that has watering holes that fill up when it rains (see Figure 6.2). Some cattle are let out to graze on and around the riverbanks where lower river flows are preventing the seeding and growth of new cottonwood trees as explained above. Lower-than-normal winter temperatures have killed or damaged the vegetation with severe frosts, translating into less winter feed for the cattle and more soil erosion. Drought has also killed some of the wild plants and trees in the *monte*. New invasive plants toxic to cattle have also been found recently in the scrublands (stakeholder workshop, interview, January 2014). For a few months, some ranchers let their cattle graze in a communally managed pasture sown with buffel grass, however, dairy producers and ranchers reported that even this drought and frost-tolerant grass has been affected, especially in terms of growth rates.

Some of the adaptation strategies to address these negative impacts of climate and precipitation may not be sustainable. Ecological tipping points are useful analytical tools for examining long-term viability of natural resources for use. Social-ecological systems (SES) are nested, multilevel systems offering essential

Figure 6.2 Cattle in the *monte* or scrublands, Rayón

Photo credit: América Lutz

services for society such as food, potable water, energy, and fiber (Berkes and Folke 1998). SES frameworks include an examination of thresholds or tipping points to shed light on linkages between human society and ecosystems (Ostrom 2007; Scheffer et al. 2002; Scott and Buechler 2013). Previously, 'tip point' had been utilized for studying social phenomena such as the point at which the influx of non-white residents into an urban neighborhood would induce 'white flight' (Grodzins 1957). The expression 'tipping point' was later popularized in a best-selling book, *The Tipping Point: How Little Things Can Make a Big Difference* (Gladwell 2000), to represent the point in time when a new idea takes root and is disseminated widely (Marten 2005). Gerald Marten then developed the concept 'environmental tipping point' in 2005 to include both positive and negative factors that precipitate significant change:

> A "negative tip" is a switch from a desirable stability domain to an undesirable one, a change from a sustainable environmental support system to deterioration of the system's services. A "positive tip" is a switch from an undesirable stability domain to a desirable one. Deterioration is turned around and the human-environment system heads toward greater sustainability.
>
> (2005, 76)

Linking feminist political ecology and SES frameworks helps to emphasize the importance of including a close examination of, in this context in Sonora, the delicate balance between gendered human activity like ranching and dairy production and ecological systems such as the hydrological cycle, desert vegetation, and vegetation that only exists near rivers.

Arid and semi-arid lands, such as those found in Rayón, are very vulnerable to desertification from climate change processes compounding negative effects of land-use patterns. In Rayón, Carla,[11] a cheese maker, said with a sigh:

> With the little rain that fell in September [2013] we are feeding our cows but if it does not rain in the coming year, I don't think that there will be cows . . . we would have to sell them. I hope it rains.
>
> (field interviews, September 2013)

Marta, another cheese maker explained, "right now there is little feed [for the cattle] because it has not rained enough" (field interviews, September 2013). Three years ago their dug well near the San Miguel River stopped giving water and the ten members of the well association had to have a deeper one drilled. A cattle rancher on a neighboring plot to theirs drilled a well but did not find water. The cost of drilling a new well is now about 100,000 pesos (U.S.$7,500), but SAGARPA (Mexico's federal agency, the Secretariat of Agriculture, Livestock, Rural Development, Fisheries and Food) will pay half of this cost to reposition a well that does not give water (as mentioned above, no new wells are allowed to be drilled) and a municipal government program, FONAVI, will pay the full labor costs. Half of this cost would be significant for most ranchers, however, even if they shouldered it as a group. One couple who are both cheese producers explained that their 40-foot-deep well stopped supplying irrigation water in March 2014, and they would like to deepen it to 100 feet, the depth at which other ranchers in the community have struck water (field interviews, March 2014). Out of eight new wells for agriculture drilled by ranchers in the end of 2013, however, only one was reported to be producing sufficient irrigation water (stakeholder meeting, Rayón, January 2014).

Ranchers are changing their herd sizes in response to drought and rising temperatures. Guadalupe, who, with the death of her husband, took over their ranch and dairy business, revealed that her cattle now get many illnesses because they are weaker due to poorer nutrition caused by lack of water to grow sufficient amounts of the wild plants and grasses that the cattle eat on her land. In early September 2013, she had to sell half of her cows due to low rainfall and because they were losing weight. She now owns 40 cows, ten calves, two bulls, and three horses. This example illustrates a case in which the ecological balance was tipped in a direction that militated against keeping cattle in the *monte*.

Dairy producers are also changing the type of cattle they purchase due to rising temperatures and prolonged drought. The Holstein cows are originally from New Zealand and do not withstand high temperatures. More ranchers are changing to Brangu (a cross between Angus and Red Zebu), which can better withstand heat and drought. Though these cows provide good meat, they produce less milk than Holsteins. Most dairy producers use milking machines and many have hired labor (predominantly men) to operate these machines. The average dairy herd size varies greatly between households, ranging from one cow to approximately 200.

Ecological knowledge is gendered in Rayón. Most ranching, including milking and agricultural production as described above, is undertaken or overseen by male household members. The case of Guadalupe, however, who took over the ranching following her husband's death, reminds us that under certain circumstances like widowhood, gender roles and responsibilities (thus ecological knowledge) can be altered. Many of the agricultural processing enterprises consist of collaborations between sisters, daughters, and mothers. Several of the women interviewed stated their reason for producing cheese thus: "there is no employment here, so one has to be self-employed" (field interviews, June 2013). Male household members sometimes assist in the transportation of milk to the home for cheese production or the transportation of cheese to market. The main type of cheese produced is *queso cocido* (cooked cheese), a semi-soft smooth white cheese produced from cow's milk used in quesadilla production (see Figure 6.3), while a minor part of cheese production is *queso fresco* (fresh cheese) from unheated cow's milk. The process of converting milk into cheese serves to preserve milk in a form that, if kept under refrigeration (or kept frozen as the cheese makers do before it is marketed), lasts a great deal longer than if it was retained in the form of milk. Women have innovated by adding different ingredients to the cooked cheese like chili peppers, dried beef, and mushrooms.

Marketing of the *queso cocido* has become increasingly diversified over time, expanding in scope to include towns and cities located some distance from

Figure 6.3 Woman cheese producer, Rayón
Photo credit: Hans Buechler

Rayón. Today, this cheese is sold to neighbors, to tourists from Hermosillo and elsewhere who visit Rayón, and to relatives of people living in Rayón who come for holidays from all over Mexico and the U.S. The cheese is also sold to itinerant vendors who operate in Hermosillo, or in Carbó and Ures (towns located about 45 minutes from Rayón) who often carry small coolers filled with ice and frozen cheese in the main plazas. In Hermosillo, the well-known and popular *queso Rayón* (cooked cheese) is often sold from urban residences (indicated by a sign in the window). The seller is often a relative of a Rayón cheese producer and is usually a woman.

Rayón cheese also acts as an important conduit in important social networks. Women serve the cheese, usually melted in *quesadillas* to migrant relatives who are home to visit. With high migration rates to Hermosillo, to other parts of Mexico, and across the western and southwestern U.S., the cheese produced and served by Rayón women creates a highly favorable social context. As one visiting migrant told me: "Visiting Rayón without eating cheese and bringing some back with me is like not going to Rayón at all" (interview, March 2012), but women and men complained that fewer people were visiting Rayón because of less water in the landscape (stakeholder workshop, January 2014).

The Cattle Ranchers' Association of the municipality of Rayón has 279 members; approximately 53.8 percent of whom are either *ejido* members or *comuneros/as* (Cattle Ranchers' Association 2012 in Moreno Vasquez and Navarro Navarro 2012).[12] This is a reduction from approximately 387 ranchers in 2004 (Plan Municipal de Desarrollo de Rayón 2004). This association makes cattle feed—mainly in the form of alfalfa bales—cheaper for the ranchers (13 pesos instead of 19 pesos); the price is subsidized by SAGARPA. These alfalfa bales come from more water-abundant areas in Sonora and Sinaloa and represent a virtual water (water used in the production of the product) transfer between regions. If those alfalfa-producing areas are not able to grow alfalfa in the future because of declining groundwater levels, they may stop exporting alfalfa to Rayón and other areas. A lack of alfalfa for dairy cattle or a steep rise in its price could impact the amount of milk available for cheese production.

Guadalupe, one of only two women members of this cattle rancher's association, asserted that when she speaks at a meeting, she notices that at the next meeting the male members ignore her. She concluded that "it is better to just keep quiet." She also relayed that it is very hard for her in Rayón because she cannot really say what she thinks since everyone is related to one another. The men I talked to thought that Guadalupe should rent out her ranch to a man and dedicate herself entirely to running her general goods store/*quesadilla* eatery. Her daughters, studying in Hermosillo and Ures, were taught by their mother, but will likely not go into dairy production or cheese making. They warned her that if she does rent out her land, then she should not rent it to a relative, because male relatives would be unlikely to pay her the rent they owe. Guadalupe does not want to sell her land to anyone. She told me with strong determination in her voice and in the set of her jaw: "I feel a sense of relaxation when I work at the ranch that I do not feel when I am in town" (interview, March 2013). The ranch is where she and her farm hand keep the cattle, and it is also where they make the cheese in the kitchen of her small ranch house. Her kitchen is the locus of family gatherings that are replete with cheese melted within *quesadillas*. During our interview, as we sat on the veranda overlooking the ranch, she noticed that shade obtained from palm trees near the house has left following the recent death of these trees due to drought. This ranch then, similar to most others in Rayón, was made into a place to live and to raise livestock, offering a degree of freedom for this woman who heads her ranch; it also represents a retreat for visiting family members who maintain social networks as they sit together and eat her cheese.

Guadalupe also explained the climate's effects on cheese production. She explained that variations in temperatures make a huge difference in cheese production, because both unusually cold and hot temperatures can kill cultures (bacteria) used to produce cheese. She has an in-law who also produces cheese and who gives Guadalupe culture if hers is killed. If temperature extremes worsen as predicted with climate change, more cultures will be destroyed, and women's strategies of borrowing cheese cultures from other women may be rendered ineffective. To keep the culture alive during periods of cold temperatures, Guadalupe also sometimes keeps the wood-burning stove in her ranch's small kitchen going while making the cheese. Other women have begun to add citric acid to the cheese when cooking it, which works with the culture to help acidify the milk.

Some women have received government assistance for their cheese enterprises. Guadalupe, for example, received government funding from controversial Plan Sonora Proyecta (PSP) to set up her store/*quesadilla* eatery; as mentioned, the cheese for this eatery comes from the cheese she produces in her kitchen. SAGARPA also gave her a grant five years ago that covered half the cost of constructing five small hotel rooms next to her house. She would like to construct more hotel rooms and is looking for more funding to be able to do this.

Silvia is another cheese producer in Rayón who has been producing cheese for about 15 years. She purchases the milk from Carbó, the nearest town, because their well no longer provides water and thus cannot be used to irrigate crops for cattle feed. She and her husband only have a few cows that feed on wild plants in the *monte* and in the communal plot of buffel grass. Carbó is a 45-minute drive away on a poorly maintained road. Silvia produces cheese with her female cousin; her sister was also working with her but recently found a daily wage job. Her 22-year-old son delivers the cheese to vendors in Hermosillo once a week. In 2007, Silvia obtained funding from a government program, Alianza Para el Campo, to purchase burners to produce more cheese at one time and freezers to store more cheese before selling it; she also received assistance to build an additional room in her house to produce the cheese. For both grants, she received half of the cost of labor and materials. She explained how she got the chief of the irrigation district to come to her house and look at the space that she used for cheese production. She showed him that she needed the equipment and extra space to expand her business. Silvia noted "he was in a meeting in the municipal office in Rayón, and I grabbed him and pulled him to my house to show him the large quantities of cheese I produced" (field interview, January 2013). Although she is grateful for the assistance, she also reported that the percentage of aid the government now gives has been greatly reduced.[13] Silvia now has her own packaging label that bears her first name, and her cheese has won awards and recognition in northern Mexico. She has innovated with another member in Rayón who makes dried beef and now mixes the dried beef into some of her cooked cheese, thus expanding the flavors of cheese she sells.

In winter 2013, Silvia added an email address to her business card for email orders. She also added a website on which customers can place orders. Recently she was able to afford home internet service using a new bundling with the local phone service.[14] She helped formulate a plan to develop a cooperative with six female and male members to build a dairy cattle stable, but thus far they have not found funding sources for this venture and the cooperative is only a dream. She indicated that she wishes to be able to raise her own cattle with the cooperative because prices of milk and gasoline inputs are increasing. She cannot raise the price of the cheese, however, because she fears that people would no longer buy it. Other studies have shown that the formation of agricultural producer groups like cooperatives can help women and their household members to reduce economic and environmental risks in the face of climate change (on Sonora, see Vásquez-León 2009 and Vásquez-León and Buechler n.d.; on Brazil, see Corral 2010; on Colombia, see Tovar-Restrepo 2010). Silvia's struggles related to funding the necessary infrastructure for the cooperative point to policy-related bottlenecks in grassroots initiatives for climate adaptation.

Studies on gender and climate change have shown that women are more receptive than men to green energy sources and can act as a climate change mitigating force (Aguilar 2013). This finding also resonates with Silvia's experience. She has goals to reduce her increasing energy costs through the purchase of solar panels to produce electricity to run the freezers. Although she knows of a government program that would provide 50 percent of the cost of installing panels, she lamented, "they [the government officials] did not return here . . . they only finally provided them for communities in Sonora in the mountains where there is no electricity" (interview, May 2012). By winter 2014, border security concerns also limited government agency staff visits to potential project sites in the region.

The cases of Guadalupe and Silvia show the creative responses that cheese makers in Rayón have made in the face of a changing climate and water resource landscape. Their examples also reveal the importance of understanding gendered knowledge in policy and programmatic planning related to climate change adaptation and water management in riparian areas. In particular, women cheese makers in Rayón have a keen understanding of the implications of climate and water resource variability on the final processing stages of agricultural production. They therefore act as leaders in climate-water adaptation and mitigation strategies. Women have changed the inputs for their products over time and have also expanded marketing of their cheese to wider circles in Sonora. They combine many different livelihood activities and often include other household members in the cheese-production processes or product marketing. The processed cheese is frequently made within their own home compounds allowing them to combine this work with other work such as childcare, eldercare, cooking, cleaning, etc. Despite having little representation in formal organizations such as the Cattle Ranchers' Association, some women have successfully negotiated government assistance for production equipment and additional production space.

The sustainability of the type of agricultural production to produce cattle feed is tenuous, nevertheless, because of increasing reliance on groundwater, as surface water sources like rivers are depleted. Negative ecological tipping points are visible in groundwater availability and have repercussions for dairy producers and processors. Every farming household in this study recently either had to deepen a well at significant cost or else completely ran out of well water. The well that served as the primary source of drinking water for Rayón was also deepened in 2013. Among other issues, lowering of groundwater tables and lack of surface water greatly affects milk supply for women's cheese-processing activities. Purchasing milk from elsewhere may also become too expensive as groundwater in those locations is similarly depleted.

Women in Rayón have engaged in creative and dynamic processes of adaptation to environmental change. Gendered knowledge is visible in adaptation strategies related to agricultural processing. Women engage in experimentation with cheese production, mostly carried out in their home compounds. They also practice flexible sourcing of inputs that they combine with livelihood diversification, including the education of their children to much higher levels than their own. They use their products as a source of income and food, and as a conduit for social ties critical to the maintenance of social networks that include migrants. Sharing of cheese helps ensure they will retain the social and economic support these migrants represent. Ecological tipping points in rural Sonora are becoming palpable, however, as water is imperiled by upstream water use in growing cities such as Hermosillo. Lower rainfall has impacted the land in the scrublands and along the river which serve as feeding grounds for their cattle. As Jordan et al. (2006, 13) have argued: "Effective responses to desertification will require socio-economic and policy interventions, but pacing such interventions to prevent irreversible ecological change remains a significant challenge and depends on timely forecasts of tipping points." Although the socio-environmental change processes in Rayón include more than desertification, the concept of tipping point is useful for understanding the complex environmental changes occurring in Rayón and the need for broad-based inclusion of gendered environmental-agricultural knowledge and solutions-oriented action.

Conclusions

Women are not visible as significant actors in agriculture, therefore government policies and programs are not tailored to meet their needs (Sogani 2013). As exemplified by the case of Rayón, they tend to inhabit spaces and stages of production that are less visible to government agencies. Agriculture and water-related policies and programs in Sonora, as in many regions of the world, focus on irrigated fodder production for livestock rearing to the neglect of rain-fed lands outside of cultivated areas (such as the ecologically wilder *monte*) used for grazing animals. These types of policies and programs are also dedicated to crop and livestock production (mainly carried out by men) to a greater degree than to the processing stages in agricultural production (which are mainly done by women). The important yet undervalued spaces of agricultural processing, such as cheese production, are linked with gendered knowledge. The use of a feminist political ecology lens fosters an appreciation of the importance of these spaces. Placing value on these spaces is a first step in unveiling the gendered expertise associated with agricultural activities in each location.

Climate change is evident to local residents of Rayón and includes extension of the current drought and more frequent and more prolonged periods of hotter- and colder-than-normal temperatures that kill crops, wild trees, and other plants. It has exacerbated rural-to-rural virtual water transfers in the form of animal feed cultivated in other locations and transported to Rayón. Because other locations are also suffering from climate change, the sustainability of this production of animal feed, particularly alfalfa, is also uncertain. Rural-to-urban water transfers intensify as urban populations grow, a growth fueled in part by economic challenges such as the declining viability of agriculture.

Gender dimensions of riparian livelihoods are increasingly important to understand in the context of ever-harsher environmental conditions in communities located along disappearing rivers. These livelihoods require increasingly complex adaptations and specialized knowledge. A feminist political ecology lens can help focus attention on expertise women possess in the multi-dimensional and gendered spaces they occupy. Women in Rayón have adapted through various means, such as experimentation with the sources of inputs, the expansion of the types and volumes of products made, and the development of creative and modernized marketing. A feminist political ecology approach focused on gendered environmental knowledge also guides attention to the inter-generational sharing of women's ecological knowledge, although currently there is insufficient attention in feminist political ecology research on this topic of knowledge transfer. Although some women in Rayón are passing their agricultural-ecological knowledge down to younger family and community members, many find that faced with rising climate challenges, young women and men are more often choosing other careers, often in urban locations, that do not include food production or processing. Feminist political ecology calls for a linking of the political and the social with ecological processes. Unlike political ecology, it lends a gender focus to these issues. By raising their voices, women in Rayón were able to obtain some government assistance to expand their processing businesses or to reduce their economic risks by diversifying household livelihood strategies. They reported, though, that many of the previous sources of government assistance for these ventures have disappeared or been reduced. These policy issues will have social ramifications such as possible negative ramifications for the social cohesion fostered by the production, marketing, and consumption of cheese.

Studies on gender, water, and climate change would benefit from a combined analytical approach utilizing political ecology and SES frameworks. Feminist political ecology and political ecology research often underplay the ecological and policy implications of the research (Forsyth 2013; Walker 2005); SES frameworks could help to influence future directions for feminist political ecology and political ecology research and analysis by focusing on the delicate ecological balance in which gendered human activities occur and by incorporating this into policymaking. Policy guided by feminist political ecology would start with an appreciation of the critical stage in agriculture that is food processing (which in Rayón is cheese production), and the actors who engage in this production. Governmental and non-governmental projects could help both mitigate climate change by, for example, encouraging and subsidizing solar energy use and by helping processors adapt to new ecological conditions such as higher temperatures and less water for agriculture. Policies could be dedicated to providing platforms for cheese makers around Mexico to share their experiences in ameliorating the negative effects of environmental change and could be combined with current expositions of food processing. Policy could also be tailored to assist female family members of cheese makers to develop other income-generating activities that act in similar ways to cheese production in maintaining household and community-level social ties and food security, but that are more sheltered from longer-term climatic and water variability. Responses such as raising very small livestock near or within home compounds and fed with household food scraps and crop residues could be one such activity. Any successful adaptive responses, as Smit and Skinner (2002) remind us, are undertaken in concert with changes at various levels (micro to macro level) with actors from households to national and international organizations to government agencies.

Utilizing a SES lens, ecological tipping points become visible in Rayón. These include groundwater depletion which will impact the nature of women's and men's participation in certain processing activities. Groundwater depletion will also impact the long-term, region-wide viability of agricultural production that underpins these processing enterprises. If water resource availability declines further in this and similar riparian communities, dairy production and therefore dairy-processing enterprises could be rendered insolvent and these gendered rural livelihoods severely impacted. As climate change impacts intensify, youth and even older community members may increase their rate of exit from these types of production and processing activities, which may in turn increase out-migration rates especially from villages such as Rayón. If officials and decision-makers do not place priority on multi-faceted agricultural livelihoods and the socially differentiated actors who engage in them, food production may further dwindle, just as the world requires additional sources and volumes of food to feed its growing population.

Notes

1 The study in this community in Sonora, Mexico was made possible by the following grant: National Science Foundation Dynamics of Coupled Human-Natural Systems grant (DEB-1010495) "Strengthening Resilience of Arid Region Riparian Corridors: Ecohydrology and Decision-Making in the Sonora and San Pedro Watersheds" (2012–2015).

2 See report for Mexico available on Aquastat, FAO's information system on agriculture and water, www.fao.org/nr/water/aquastat/countries_regions/MEX/index.stm. Accessed May 21, 2014.

3 For information about COTAS see www.cotassonora.com. Accessed June 5, 2014.
4 CONAGUA is the technical advisory commission of Mexico's Ministry of the Environment and Natural Resources, SEMARNAT.
5 Additional information provided by COTAS staff (personal communication, January 2014) and COTAS website, www.cotassonora.com. Accessed April 29, 2014.
6 See Censo de población y vivienda 2010, Consulta rápida, available from the Instituto Nacional de Estadística y Geografía (INEGI), www3.inegi.org.mx/sistemas/ResultadosR/CPV/Default.aspx?texto=Hermosillo. Accessed May 3, 2014.
7 Sex-disaggregated statistics are currently being obtained via household surveys on agriculture, livestock, and fishing livelihoods and on food security for other regions of the world such as sub-Saharan Africa (IUCN 2013, 55).
8 See statistics for Mexico at www.fao.org/nr/water/aquastat/countries_regions/MEX/index.stm and map of irrigation areas in Mexico at www.fao.org/nr/water/aquastat/irrigationmap/mex/index.stm. Accessed June 8, 2014.
9 See Censo de población y vivienda 2010 available from INEGI at www3.inegi.org.mx/sistemas/ResultadosR/CPV/Default.aspx?texto=Hermosillo. Accessed May 3, 2014.
10 An *ejido* is a community with communal agricultural land, with each member farming an individual plot. This information comes from the Plan Municipal de Desarrollo, 2003–2006, from the Presidencia Municipal de Rayón, Sonora, and from interviews by the author in Rayón, Sonora, January 2012–January 2014.
11 Pseudonyms are used throughout this chapter to help protect the identity of the interviewees.
12 A *comunero/a* is a member of an area of communally held land that may also, as in the case of Rayón, have *ejido* land (see note 10) and is also part of the 'social sector' of non-private landholders with use rights to the land. Neoliberal reforms to Mexico's constitution and to agrarian legislation in 1992 legalized the sale of both types of landholdings in this 'social sector.' For a history of the land tenure system in Mexico see Assies (2008).
13 Declining government funding for social programs is due in part to misuse of federal government funds by state and municipal governments (Ortega and Fraijo 2014).
14 It is possible that some dairy producers will be able soon to make use of their easier access to the internet to obtain climate and rainfall information that could be of assistance to them in their decision-making. By winter 2014 residents in Rayón mainly gained access to the internet through their cell phones but some also had computers. Government data on computer ownership in Rayón are from 2010 and with the dynamism in this area, the statistics are very outdated.

References

Aguilar, L. 2013. A path to implementation: Gender-responsive climate change strategies. In *Research, action and policy: Addressing the gendered impacts of climate change*, eds M. Alston and K. Whittenbury, 149–157. New York: Springer.

Assies, W. 2008. Land tenure and tenure regimes in Mexico: An overview. *Journal of Agrarian Change* 8(1): 33–63.

Barndt, D. 2002. *Tangled routes: Women, work and globalization on the tomato trail*. Lanham, MD: Rowman & Littlefield.

Berkes, F. and C. Folke, eds. 1998. *Linking sociological and ecological systems: Management practices and social mechanisms for building resilience*. New York: Cambridge University Press.

Buechler, S. 2009. Gender, water and climate change in Sonora, Mexico: Implications for policies and programmes on agricultural income generation. Special issue on climate change. *Gender and Development* 17: 51–66.

——. 2012. Sustainability of gendered agricultural production in the context of water scarcity and climate change on the Mexican border with the United States. In *Gender and sustainability: Critical case studies from Asia and Latin America*, eds M.L. Cruz Torres and P. McElwee, 121–141. Tucson, AZ: University of Arizona Press.

Cabrera, I.R., E.Z. Martelo, and V.V. García. 2001. Gender, rural households, and biodiversity in native Mexico. *Agriculture and Human Values* 18: 85–93.

Carney, J. 2001. *Black rice: The African origins of rice cultivation in the Americas*. Cambridge, MA: Harvard University Press.

CEA Sonora (Comisión Estatal del Agua en Sonora). 2008. *Estadísticas del agua en el estado de Sonora*. Hermosillo: CEA Sonora.

CONAGUA (Comisión Nacional del Agua). 2009. *Actualización de la disponibilidad media anual de agua subterránea. Acuífero (2625) Rio San Miguel. Estado de Sonora*. Mexico: CONAGUA.

Cook, B.I. and R. Seager. 2013. The response of the North American monsoon to increased greenhouse gas forcing. *Journal of Geophysical Research* 118: 1690–1699.

Corral, T. 2010. Gender perspectives in adaptation strategies: The case of pintadas solar in the semi-arid region of Brazil. In *Gender and climate change: An introduction*, ed. I. Dankelman, 138–144. London: Earthscan.

Elmhirst, R. 2011. Introducing new feminist political ecologies. *Geoforum* 42: 129–132.

Forsyth, T. 2013. Community-based adaptation: A review of past and future challenges. *Wiley Interdisciplinary Reviews: Climate Change* 4(5): 439–446.

Garfin, G., A. Jardine, R. Merideth, M. Black, and S. LeRoy, eds. 2013. *Assessment of climate change in the Southwest United States: A report prepared for the National Climate Assessment*. Washington, DC: Island Press.

Gladwell, M. 2000. *The tipping point: how little things can make a big difference*. Boston, MA: Little, Brown.

Grodzins, M. 1957. Metropolitan segregation. *Scientific American* 197: 33–47.

Gunewardena, N. 2010. Bitter cane: Gendered fields of power in Sri Lanka's sugar economy. *Signs* 35: 371–396.

Hanjra, M.A. and M.E. Qureshi. 2010. Global water crisis and future food security in an era of climate change. *Food Policy* 35(5): 365–377.

House-Peters, L. and D. Chan. 2014. Strengthening the resilience of arid region riparian corridors. Presentation to Coupled Natural-Human Systems NSF project 'Strengthening the resilience of arid region riparian corridors' second stakeholder workshop. Sierra Vista, AZ, February 12.

Hough, M., C. Scott, M. Pavao-Zuckerman. n.d. From plant traits to social perception: A social-ecological systems approach to identifying thresholds in ecosystem services. Manuscript in preparation.

INEGI (Instituto Nacional de Estadística y Geografía). 2011. Censo Bovino. Produccion Mensual para el Año Arícola 2011. Mexico: INEGI.

IUCN (International Union for Conservation of Nature). 2013. *The environment and gender index (EGI) 2013 pilot*. Washington, DC: IUCN. http://environmentgenderindex.org/. Accessed March 25, 2013.

Jackson, R.B., S.R. Carpenter, C.N. Dahm, D.M. McKnight, R.J. Naiman, S.L. Postel, and S.W. Running. 2001. Water in a changing world. Issues in ecology No. 9. Washington, DC: Ecological Society of America.

Jordan, T.E., A.H. Lloyd, J.W. McClelland, C. Langdon, D.A. Mouat, K.M. Havstad, and J.A. MacMahon. 2006. Ecological tipping points: Subtle alterations may signal the approach to drastic transformations of ecosystems affected by global climate change. Paper based on the November 16, 2006 briefing and symposium sponsored by AERC, Washington, DC. www.ecosystemresearch.org/tipping%20points%20ms%20%202%2013%2007.pdf. Accessed May 30, 2014.

Lankao, P. and J.B. Smith (coordinating authors). 2014. Chapter 26. North America. Intergovernmental Panel on Climate Change (IPCC). Assessment report 5. http://ipcc-wg2.gov/AR5/images/uploads/WGIIAR5-Chap26_FGDall.pdf. Accessed April 22, 2014.

MacDonald, G. 2010. Water, climate change, and sustainability in the Southwest. *Proceedings of the National Academy of Sciences* 107: 21256–21262.

Marten, G.G. 2005. Environmental tipping points: A new paradigm for restoring ecological security. *Journal of Policy Studies* 20: 75–87.

Momsen, J. 2009. *Gender and development*. 2nd ed. London: Routledge.

Moreno Vasquez, J.L. and L.A. Navarro Navarro. 2012. El fortalecimiento de la resilencia de los corredores riparios aridos: Ecohidrología y toma de decisiones en la cuenca del Río San Miguel. Project progress report. Hermosillo: El Colegio de Sonora.

Nightingale, A. 2011. Bounding difference: Intersectionality and the material production of gender, caste, class and environment in Nepal. *Geoforum* 42: 153–162.

Oakley, E. and J.H. Momsen 2007. Women and seed management: A study of two villages in Bangladesh. *Singapore Journal of Tropical Geography* 28(1): 90–106

Ortega, J. and C. Fraijo. 2014. ¡Misterio, deuda de Sonora! [Mysterious Sonoran Debt!]. DossierPolitico.com, February 27. www.dossierpolitico.com/vernoticiasanteriores.php?artid=140040&relacion=dossierpolitico&categoria=&criterio=ife. Accessed March 28, 2014.

Ostrom, E. 2007. A diagnostic approach for going beyond panaceas. *Proceedings of the National Academy of Sciences* 104: 15181–15187.

Padilla Calderón, E. 2012. *Agua, poder y escaséz. La construcción social de un territorio en un ejido Sonorense, 1938–1955*. Hermosillo: El Colegio de Sonora.

Parasecoli, F. 2010. The gender of geographical indications: Women, place, and the marketing of identities. *Cultural Studies ↔ Critical Methodologies* 10: 467–478.

Plan Municipal de Desarrollo de Rayón. 2004. Work plan. President of the Rayón municipality.

Sachs, C.E., K. Gajurel, and M. Bianco. 1997. Gender, seeds and biodiversity. In *Women working in the environment*, ed. C.E. Sachs, 177–191. Washington, DC: Taylor & Francis.

Scheffer, M., F. Westley, W.A. Brock, and M. Holmgren. 2002. Dynamic interaction of societies and ecosystems – linking theories from ecology, economy, and sociology. In *Panarchy: Understanding transformations in human and natural systems*, eds L.H. Gunderson and C.S. Holling, 195–239. Washington, DC: Island Press.

Scott, C.A. and S. Buechler. 2013. Environmental and social change in the U.S. Mexico borderlands: Climate, water, and demographic futures. *Ecosphere* 4: 1–16.

Smit, B. and M. Skinner. 2002. Adaptation options in agriculture to climate change: A typology. *Mitigation and Adaptation Practices and Policies for Global Change* 7(1): 85–114.

Sogani, R. 2013. Climate change: A Himalayan perspective "Local knowledge – The way Forward." In *Research, action and policy: Addressing the gendered impacts of climate change*, eds M. Alston and K. Whittenbury, 265–275. New York: Springer.

Tovar-Restrepo, M. 2010. Climate change and indigenous women in Colombia. In *Gender and climate change: An introduction*, ed. I. Dankelman, 145–151. London: Earthscan.

Vásquez-León, M. 2009. One decade of drought and two of neoliberal reforms in the Sierra Sonorense: Responses by the rural poor. *Southern Rural Sociology* 24: 44–66.

—— and S. Buechler. n.d. Cross-border social networks, gender and climate vulnerability in rural northwest Mexico.

——, C.T. West, and T.J. Finan. 2003. A comparative assessment of climate vulnerability: Agriculture and ranching on both sides of the U.S.–Mexico border. *Global Environmental Change* 13: 159–173.

Vivoni, E., C.J. Watts, J.C. Rodriguez, J. Garatuza-Payan, L.A. Mendez-Barroso, E.A. Yepez, J. Saiz-Hernandez, and D.J. Gochis. 2007a. Relation between surface flux measurements and hydrologic conditions in a subtropical scrubland during the North American monsoon. Presentation for the American Geophysical Union fall meeting. http://vivoni.asu.edu/pdf/AGU2007SonoraTower.pdf. Accessed August 10, 2013.

——, H.A. Gutierrez-Jurado, C.A. Aragon, L.A. Mendez-Barroso, A.J. Rinehart, R.L. Wyckoff, J.C. Rodriguez, et al. 2007b. Variation of hydrometeorological conditions along a topographic transect in northwestern Mexico during the North American monsoon. *Journal of Climate* 20: 1792–1809.

Walker, P.A. 2005. Political ecology: Where is the ecology? *Progress in Human Geography* 29(1): 73–82.

Welsh, L.W., J. Endter-Wada, R. Downard, and K.M. Kettenring. 2013. Developing adaptive capacity to droughts: The rationality of locality. *Ecology and Society* 18: 7pp.

Wilcox, B.P. and Y. Huang. 2010. Woody plant encroachment paradox: Rivers rebound as degraded grasslands convert to woodlands. *Geophysical Research Letters* 37(7): L07402. doi:10.1029/2009GL041929.

Wilder, M. 2008. Equity and water in Mexico's changing institutional landscape. In *Water, place, and equity: Apportioning water among places and values*, eds R. Perry, H. Ingram, and J. Whiteley, 95–116. Cambridge, MA: MIT Press.

——, C.A. Scott, N. Pineda Pablos, R.G. Varady, and J. McEvoy. 2010. Adapting across boundaries: Climate change, social learning, and resilience in the U.S.–Mexico border region. *Annals of the Association of American Geographers* 100: 917–928.

7 International partnerships of women for sustainable watershed governance in times of climate change

Patricia E. (Ellie) Perkins and Patricia Figueiredo Walker

Introduction

This chapter describes and assesses collaborative research with women actively engaged in local and global community engagement processes for water management in times of global climate change. As an equity-focused response to climate change, the interrelated networks and initiatives described in this chapter involve organizations and individuals in Brazil, Mozambique, South Africa, Kenya, and Canada.[1] These collaborations are focused on strengthening low-income women's voices, and legitimizing their knowledge and action within water management institutions and processes.

The theoretical framework for this analysis links the three main themes of feminist political ecology, as described in Dianne Rocheleau's foundational book: gendered environmental *knowledges*, *rights and responsibilities*, and *politics and grassroots activism* (Rocheleau et al. 1996; Elmhirst, 2011). It also reflects what Ariel Salleh, citing Maria Mies and Veronika Bennholdt-Thomsen, calls "the methodology of working with the 'view from below' . . . (which represents) capacity building for the global North" (Salleh 2009, 292, 304). As Salleh notes:

> It is surely a good time now, for professionals and global justice activists to sit down and talk together, and for both to talk with people who have a developed capacity for eco-sufficiency. But in this, there is a respectful caveat to observe too; as Australian Aboriginal activist Lilla Watson put it: If you have come to help me, you are wasting your time. But if you have come because your liberation is bound up with mine, then let us work together.
>
> (2009, 307–308)

Rather than emphasizing the theoretical and academic contributions of feminist political ecology, this chapter thus focuses on collaborative local and global participatory methods for building social-environmental change—the "how to" of grassroots feminist political ecology.

Importance of women's knowledge and engagement

The initiatives discussed in this chapter provide some examples of how women's experiences and knowledge relate to watershed governance and to global climate change. In a feminist political ecology sense, these stories link gendered environmental knowledge with rights, responsibilities, politics, and grassroots activism at various scales, both within and across watersheds.

Watersheds are examples of a fractal organization, which occurs throughout nature (veins in leaves, circulatory systems, etc.), where patterns at small scales are repeated at larger scales, and interrelationships across scales emerge organically. In the same way, and through this collaborative research, we argue that women's expertise and understanding of social and ecological processes at the local, household, and inter-household scales provide the basic knowledge and insights for watershed management that can be applied at larger (national and international) scales.

Women have special contributions to make towards watershed governance and water-related climate change adaptation because of gendered differences in their positional knowledge of ecological and water-related conditions. Although women fill multiple roles at home as well as in local communities (and of course gender combines with other aspects of identity including ethnicity, class, and race in affecting every person's political position), women in general are underrepresented, and in many cases excluded, from political and environmental decision-making processes. The dominant climate change narrative often presents women as "victims, rather than as agents capable of contributing to solutions" (Terry 2009, 3). However, in rural areas throughout Asia, Africa, and in many parts of Latin America, women are the principal managers of natural resources, which they and their families rely on for their livelihoods. More and more, climate change and climate variability are negatively impacting women's everyday lives and their ability to perform these daily tasks. "Women and women-headed households are particularly vulnerable to the combined impacts of food price rises, insecurity, and changes to climate" (Toulmin 2010, 150).

Urban dwellers may be even more vulnerable to extreme weather events than those living in rural areas, and again women are most affected. In particular, those living in low-income, dense, and flood-prone slums are experiencing more intense and longer floods and landslides. In many countries in the Global South, the increasing frequency and magnitude of extreme weather events may cause even more socio-ecological disasters in both urban and rural areas (Mirza 2003). These disproportionately impact women, who typically possess fewer financial and social resources than men and are therefore more vulnerable to the adverse impacts of climate change. Both at work and at home, "as gendered work and family responsibilities make poor women the main cleaners and caregivers, poor women are the ones most affected by water issues" (Moraes and Perkins 2007, 486). For example, a recent study found that natural disasters on average kill more women than men or kill women at an earlier age, due to their "everyday socio-economic status" (Neumayer and Plümper 2007, 551).

This indicates that women's vulnerability to climate change and variability is strongly influenced by their gender roles.

> Interventions that create greater awareness and understanding of the complex links between gender equality and the environment can help to build the capacity of the poor, especially poor women, to adapt to the impacts of, and take action on climate change.
>
> (CIDA 2002, 3)

Education and involvement of women in formal decision-making processes can strengthen their adaptive capacity. This also furthers climate justice, which involves the principles of avoiding dangerous climate change, forward-looking responsibility, putting the most vulnerable first, and fair participation of all (Paavola et al. 2006).

A growing number of grassroots climate change initiatives focus on laying the groundwork for broader political participation, which can facilitate long-term, self-directed, and community-based climate change defense strategies (Ensor and Berger 2009). These locally based approaches promote capacity-building, community empowerment, social inclusiveness, and participation. There is an extensive literature on the theoretical and practical grounding for such participatory processes (e.g. Faucheux and Hue 2001; Holland 1997; Perkins 2003; Sagoff 1998; van den Hove 2000; Wilson and Howarth 2002).

Bottom-up climate change organizing:

> has been seen as the cornerstone of an inclusive/deliberative approach to planning and governance that places stakeholders' knowledge, opinions and aspirations at the centre of decision-making, as opposed to a managerialist (technical–rational) approach in which professional expertise and bureaucratic control shape policy and practice. Participation has been promoted both instrumentally, as a "means" of ensuring that decisions are better geared toward their objectives, and as an empowering "end" in itself, ceding communities greater control over the decisions that affect their lives.
>
> (Few et al. 2007, 48)

In contrast to top-down policies such as international carbon trading or REDD (Reducing Emissions from Deforestation and Degradation) which require elaborate international rules and institutional structures, bottom-up approaches encourage communities to use local skills and knowledge to identify their vulnerabilities to climate change and develop adaptation strategies that are tailored to their needs. Building women's adaptive capacity, in particular—besides assisting individual women to become politically active and develop their leadership potential—can contribute to enhancing the adaptive capacity of communities, given the significant roles women hold (e.g. for water provision and food production).

As Ensor and Berger point out, climate adaptation should reduce women's vulnerability, which helps to meet both climate justice and broader development goals (2009, 16). Moreover, climate change education, community organizing, and leadership development increase the "voice" of marginalized people as part of a global movement to address global inequities and the fundamental drivers of climate change. This goes far beyond "adaptation" to include broader political action as well. As feminist political ecologists demonstrate (see also Salleh 2009), when women share their own environmental knowledge, this often leads them to make human rights-based claims for political rights, which inevitably changes politics more broadly.

Community-based education and organizing are fundamental to creating the conditions for local knowledge to be shared and utilized, through equitable democratic participation. Building inclusive governance structures and strengthening the role of civil society, especially women, addresses vulnerability and fosters resilience and sustainability in urban centers as well as rural areas. Non-governmental organizations in the Global South, and in marginalized areas of the North, have expertise on how to do this. Their knowledge is shareable and inspirational, and can serve as the basis for a climate change intervention approach that is progressive, constructive, and democratic. The best local initiatives are relevant and interesting for local residents and increase their job opportunities, knowledge of watershed issues, understanding of basic political and ecological principles, and confidence to express and act on their views. This community action, in turn, increases the resilience and sustainability of watershed and climate change decision-making processes. It also lays the groundwork for community organizing and extension of the environmental education activities to larger constituencies in other areas affected by climate change.

For examples of how this works, we will draw from what we learned through two international projects—the Sister Watersheds project with Canadian and Brazilian partners (2002–2008), and a Climate Change Adaptation in Africa project with partners in Canada, Kenya, Mozambique, and South Africa (2010–2012). Both demonstrated the wide applicability of local-level efforts in vulnerable communities in Toronto and in African and Brazilian cities to address equity challenges by developing strategies and materials to increase the knowledge, interest, and engagement of local residents on water-related and climate change issues, focusing in particular on women and youth.

Sister Watersheds: education for equitable water governance in Brazil and Canada

As in many countries, deforestation and environmental degradation, rural-urban migration, and continued urbanization exacerbate Brazil's vulnerability to climate change. New weather patterns have caused periodic flooding in Rio de Janeiro and São Paulo, damage to housing and infrastructure, and deaths—more than 900 in January 2011 and nearly 400 in March 2013.

While Brazil has a progressive watershed management system, requiring participation by civil society representatives on watershed committees, low-income people and women in particular are underrepresented. In 2005, women made up less than 25 percent of watershed committee members (Moraes and Perkins 2007, 488–489). Watershed committees are formed "so that water users can collectively help to decide issues of allocation, infrastructure and regulation at the watershed level" (Perkins 2004, 8; Hinchcliffe et.al. 1999). However, social norms and perceptions, overwork, and illiteracy all affect women's ability to participate in these committees. For example, if water management is presented as a technical issue instead of a social one, women may feel and actually be excluded from policy discussions. But special training and organizing can support women to participate actively and effectively in articulating their views, insights, and perspectives on local climate change and water-related challenges.

Clearly the challenges of how to elicit and motivate women's effective public involvement among economically stressed groups in civil society, how to easily convey and discuss complex ecological issues in public workshops, and how to build bridges between disadvantaged local communities and public officials at the watershed level know no boundaries: this is a global problem.

The Sister Watersheds project (2002–2008) linked universities and NGOs in Canada and Brazil in developing strategies and materials for increasing the knowledge, interest, and engagement of local residents on water-related issues, focusing on low-income neighborhoods in São Paulo and Toronto, and in particular on low-income women. This C$1.3 million project—funded by the Canadian International Development Agency (CIDA) through the Association of Universities and Colleges of Canada—combined student exchanges, research, community engagement, and "capacity-building" in local communities and nearby universities. Its novel conceptualization and design were developed by progressive Brazilian environmental educators—Dr Marcos Sorrentino (a professor at the University of São Paulo who was subsequently appointed director of environmental education in the Brazilian Federal Ministry of the Environment) and Larissa da Costa of the Ecoar Institute for Citizenship (Ecoar), who subsequently became environmental education director at the World Wildlife Fund in Brasilia. The project's design evolved throughout its implementation by organizers at Ecoar, a leading environmental education NGO based in São Paulo, and at York University in Toronto, Canada.

Addressing the lack of participation by civil society representatives, especially women, in water sector governance, which of course has parallels in every country including Canada, was a primary objective of the Sister Watersheds project. The project developed and tested training programs by conducting workshops led by its local NGO partners with more than 1,450 participants, approximately two-thirds of them women, and by partnering with other community organizations on environmental and watershed management education. For example, staff from Ecoar contacted groups of elementary school teachers, public health extension agents, youth groups, and other community-based workers, and provided in-service training for them about water and health, basic ecology, and public policy questions related to water in their local communities. The various training programs were designed to be specifically appropriate for groups of women, children, youth, health agents, school groups, teachers, film/culture/music/arts organizations, and Agenda 21 groups. This helped to link non-governmental, community, and government institutions into watershed governance, by showing participants how their experience and knowledge of the watershed were important in larger political processes. The workshops focused on water management, environmental education, community development, and democratic participation, with particular emphasis on gender and socio-economic equity. The methodologies, techniques, and materials developed for these workshops and training programs—made freely available to other organizations through publications and websites—contributed to the capacity of project partner organizations, individual staff members, and students to continue their work on watershed policy issues.

The curriculum materials and techniques developed by the project were tested and fine-tuned in more than 220 workshops designed and led by project staff, student interns, and university exchange students in three watersheds—the Piracicaba and Pirajussara watersheds in São Paulo, Brazil, and the Black Creek watershed in Toronto, Canada—where university campuses are located near low-income residential areas or *favelas* (see Figure 7.1). All the project workshops were conducted with low-income residents, and the workshop methodologies were designed to build participants' confidence about their knowledge and its importance, the need for their intervention as representatives of their community, and their ability to participate in watershed governance processes. All of the Brazilian workshop participants were potential civil society participants in watershed committees, which were well-organized in their areas as a result of Brazil's 1998 water law. Some of the academic partners in the project held positions of responsibility on technical groups connected with the watershed committees, so inter-personal connections also helped to build workshop participants' engagement.

The outreach materials developed by the project include an illustrated Manual on Participatory Methodologies for Community Development containing a set of workshop activities and background materials for participatory community environmental education programs and training sessions with marginalized community members focusing on water and other environmental topics and equity issues (Teixeira et al. 2008); an illustrated guide with practical exercises focusing on urban agroecology (Gonçalves et al. 2008); a full-color socio-environmental atlas bringing together ecological, hydrological, and social information about one local watershed in a series of interactive maps which was made available to local political and watershed committee leaders; videos about the project and about the history and environment of the watersheds (Projeto Bacias Irmãs 2007a, 2007b); a publication outlining Agenda 21 activities in schools; and several blogs and websites with materials and discussion-starters on watershed topics, as well as a book and many journal articles, masters' papers, and

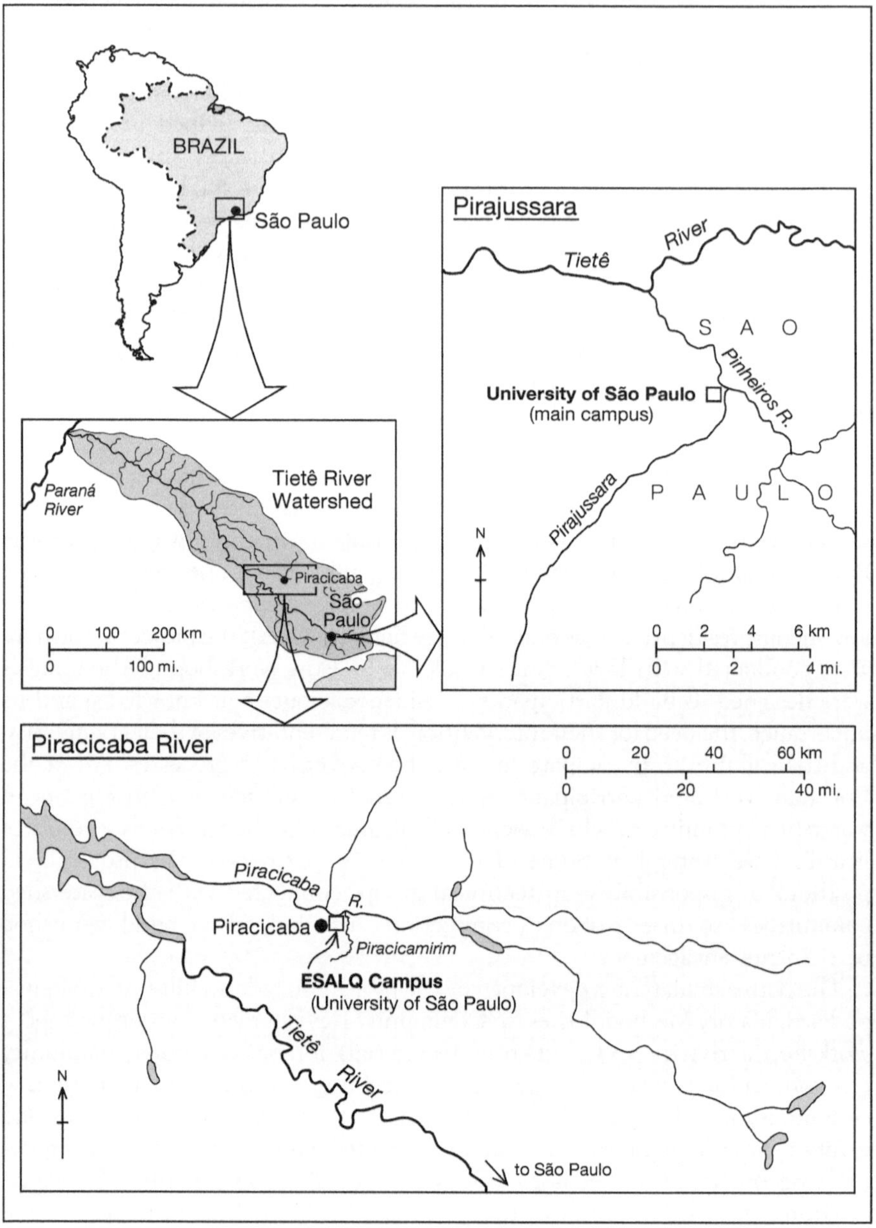

Figure 7.1 Map of São Paulo watersheds and university campuses

Source: Carolyn King, Cartographic Office, York University other academic publications contributing to literature on participatory watershed education in Brazil and in Canada (Sister Watersheds Project Final Report 2008).

A middle-aged woman from a São Paulo *favela* (slum) working with local health extension and outreach in slum areas (primarily with women and children) stated this about a Sister Watersheds environmental education workshop:

> *O rio córrego como protagonista é a pauta principalmente em área urbana, portanto, toda a discussão em torno disso é necessária e muito importante. A maneira como está sendo apresentada no curso me parece bastante satisfatória. Tem funcionado, na verdade como condutor de reflexão e discussão enriquecendo conhecimentos.* (The river as protagonist was the main agenda, principally in the city, therefore all the discussion about this is necessary and very important. The way this is presented in the course seemed very satisfactory to me. It functioned in fact like a conductor of reflections and discussion, enriching our knowledge.)
>
> (Sister Watersheds Project Final Report 2008, appendix 1)

A 14-year-old participant in a series of project workshops commented:

> *Para muitos que hoje estão aqui, esse é apenas o final do projeto, mas para mim isso é o começo. Vocês semearam em nós o respeito e o zelo para com o meio ambiente e o meu maior desejo é disseminar esse carinho e cuidado com o planeta. Obrigada por tudo e obrigada a todos que juntos estão trabalhando por um mundo melhor.* (For many of us here today, this is just the end of the project, but for me this is the beginning. You have planted in us respect and zeal for the environment, and my main desire is to spread this affection and care for the planet. Thank you for everything, and thanks to all who are working for a better world.)
>
> (Emanuela, female student in Vera Lucia Fusco Borba school, age 14—Projeto Bacias Irmãs 2008, 25)

Community environmental perception surveys conducted by the project in each of the Brazilian watersheds established a database of information on public priorities and views on watershed issues. A socio-environmental atlas gathered and made available a wide range of information on ecological, hydrological, social, and political circumstances in the watershed—information which proved very useful to public officials and watershed committee members in understanding the watershed as a whole. The nearly 1,500 (at least 942 female and 536 male) participants in workshops conducted by the project gained familiarity and experience with water-related issues and their own ability to influence water management and policy through participation in existing watershed committee structures, as well as community organizing, community arts, sharing their knowledge and care for the local streams and rivers.

This project helped both its university and NGO participants to bridge the gap between academic and community-based methods of environmental education.

Graduate exchange students studied and contributed to local training programs; faculty members wrote about the theoretical and practical benefits of public participation in watershed management; NGOs supervised students who received academic credit for their community-organizing work; professors led local watershed governance technical committees; and innovative methods for environmental education were shared internationally. This collaboration allowed new perspectives on water management to evolve, with benefits for all participants' training/education programs. The University of São Paulo, York University, and Ecoar developed dozens of new partnerships with other community organizations as a result of this project. Students, both in Brazil and in Canada, played a crucial role in developing the linkages between academic institutions and community-based NGOs. Both locally and internationally, students sought out community organizations for their research and field experiences, and shared the results of their work with both academic and non-academic audiences. The student exchanges sponsored by the project thus fuelled its interdisciplinary and educational bridging contributions. The project also showed that ongoing applied research projects and community action on watershed governance are needed, in both Brazil and Canada, to build equitable watershed-based networks of civil society, academic and government actors in order to increase women's leadership and participation in watershed-scale climate-related challenges.

The methods and approach of the Sister Watersheds project proved to be applicable to climate change education and organizing in Canada as well as in Brazil. Beginning in the summer of 2006, using some of the workshop ideas and materials generated in Brazil, several teams of York University graduate assistants and volunteers conducted a series of environmental education workshops with young people from the Jane-Finch neighborhood. This neighborhood, one of the most stigmatized in Canada and widely known for its ethnically diverse population, crime rate, and poverty as well as infrastructure and urban planning problems,[2] borders York University across Black Creek, a tributary of the Humber River (see Figure 7.2).

Working together with local organizations, students and project volunteers incorporated environmental education into existing summer-camp and after-school programs in a variety of local community centers and other venues. This programming contributed environmental content to the summer experiences of the young people and brought an environmental focus into the ongoing work of local organizations, while giving York University students practical opportunities to test their environmental education skills. The materials and workshop ideas generated for these summer programs served as the basis for Sister Watersheds graduate assistants' participation with a local community development NGO, Doorsteps Neighborhood Services, in its after-school programs for children aged eight to 13. Ecoar organizers from Brazil visited the Doorsteps programs, which enriched programming ideas, materials, and techniques on both sides. This partnership showed that there are far more international commonalities than differences in doing community-based environmental education with youth and women.

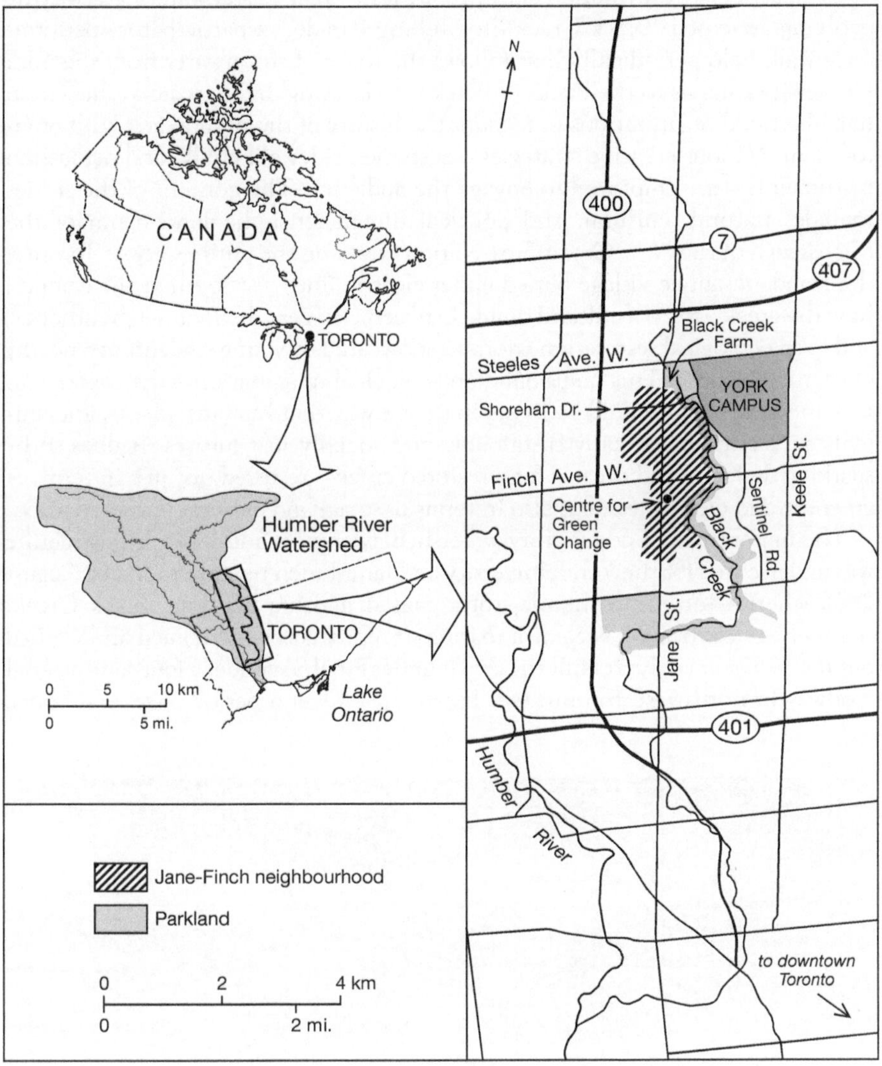

Figure 7.2 Map of Black Creek, York University, and Jane-Finch neighborhood

Source: Carolyn King, Cartographic Office, York University

Most of the neighborhood youth workshop participants were girls, since the boys tended to gravitate towards sports-centered programs, and parents in the area, largely new Canadians, were more concerned to have their daughters enrolled in organized programming. (When the environmental education included pulling on hip-waders and taking samples of the water in Black Creek for benthic invertebrate monitoring, however, the boys were very interested in joining in.) The students and staff leaders of the workshops were mainly women, by a ratio of more than four to one.

One particular contribution of the York/Black Creek "sister watershed" was the evolving art-based "Black Creek Storytelling Parade," a participatory performance walk held periodically that follows the route of stormwater from the York University campus to the banks of Black Creek, using different storytellers from neighborhood organizations to recount the history of the natural and built environment. Various creative strategies—costumes, sidewalk chalk, and percussion instruments—are employed to engage the audience. The content of the stories includes natural, cultural, and political dimensions: the land claim of the Mississauga of the New Credit First Nation covering the entire City of Toronto, the Haudenesaunee village buried under electric lines just south of the campus, how the creek acts as a cultural divide between two very distinct neighborhoods, and local ecological restoration efforts as showcased by young students at a nearby elementary school. This helps build an ecological imagination—the capacity to imagine how people could be living in such a way that humans, plants, and animals thrive in ecologically sustainable and socially just futures. It does so by sparking dialogue and dreams for a restored creek—restored not just in terms of greenery and cleanliness, but also in terms of social and cultural importance.

The importance of community watershed awareness and water infrastructure was underscored for the York University and Jane-Finch neighborhood in August 2005 when a torrential thunderstorm caused massive flooding. Black Creek, normally a slow trickle, became a roaring torrent that overwhelmed and washed out the 1950s-era culvert built to carry it under Finch Avenue, a four-lane arterial roadway in northwest Toronto (see Figure 7.3). Over a period of several hours,

Figure 7.3 Black Creek washout of Finch Avenue, 2005
Photo credit: Lucas Oleniuk/Getstock.com

the creek carved a chasm about 50 meters wide and ten meters deep. Repairs, including a new bridge for Finch Avenue, cost more than U.S.$3 million and took nearly six months to be completed. During this time, commuter traffic and city buses were diverted through the York University campus, causing major disruptions for the university and local residents alike.

In 2008, the Jane-Finch Community and Family Centre, a local social services organization, began applying for funding to develop jobs training for local residents in the area of energy retrofit evaluation, community garden development and local food production, and other "green jobs" such as insulation and renewable energy construction. The vision for this project derived from recognition that low-income people are often the most negatively affected by environmental problems, and that poverty, social justice, and the environment are interlinked.[3] The successful Green Change Project they created, with funding from various programs of the City of Toronto and the Province of Ontario as well as the United Way, an umbrella social services NGO, has won awards and is developing new models for training Green Change Agents—local Jane-Finch residents who gain skills, references, confidence, and a forward-looking perspective which increases their employment opportunities (JFCFC 2013). York University alumni, students, and faculty are involved in developing the curriculum, teaching workshop modules, and serving on the advisory board for the Green Change Project. At a gala fundraiser in March 2011, keynote speaker Majora Carter[4] overviewed similar initiatives in many other low-income neighborhoods where community development, green job creation, and ecological restoration are combining to produce powerful synergies. When community-based organizations can patch together funding through a range of government job-creation, social assistance, and environmental programs, they may be able to generate training programs and create local jobs in environmental remediation, alternative energy projects, food production/processing, and other sectors.

According to the first Green Change Project coordinator, Rosemarie Powell,

> The neighborhood of Jane and Finch is plagued with violence due to socio-economic struggles and the endemic rate of poverty and underemployment. On the other side of the coin is the emerging trend and related opportunities in the green economy. . . . Our vision is for a Centre for Green Change in the heart of Jane and Finch, where residents and youth concerned about the protection of the environment are engaged and mobilizing others as they increase their knowledge and skills and initiate individual and collective actions toward building a healthy, safe, prosperous, and environmentally friendly neighborhood.
>
> (Powell 2010, 6, 4)

The Green Change Project was developed and is led by community activists, mainly women; by June 2013, about two-thirds of the Green Change Agents trained were women. Green Change Agent Janet Campbell, who was hired to work as a local environmental educator and led meetings on the community's right-to-know about toxics emitted locally by industrial polluters, proudly said, "Thanks to the Green Change Project, I am actively involved in taking care of our planet" (JFCFC 2012, 2).

As part of a subsequent international research/linkage project, two York University graduate students wrote overviews of climate change and water-related equity issues in Toronto, in preparation for their exchange visits to African cities to study similar questions there (Lorimer 2011; Todd 2011). One of these Toronto studies was a history of the Green Change Project, which project organizers could use on their website and in future grant proposals; the other was an overview of the impacts of climate change and adaptation policies in Toronto through an equity/climate justice lens.

The 2005 Black Creek flood was a graphic example of how increasingly common extreme weather events—in conjunction with aging infrastructure, urban sprawl (including campus development), and increasingly rapid rainfall runoff due to impermeable urban surfaces—can have costly and traumatic effects on everyone in the watershed. Climate change affects all of us together! Sharing social knowledge at the watershed level greatly facilitates adaptation to such new realities. In 2014, this neighborhood-scale project was still generating local environmental education, climate adaptation, and green community development in the Jane-Finch area, bringing local community organizers and residents into collaboration with students and faculty from York University—nearly all women.

Climate change and urban water governance in Africa

Contacts among several academic and civil society organization (CSO) partners in three African countries and Canada led to the development of a new project entitled "Strengthening the role of civil society in water sector governance towards climate change adaptation in African cities—Durban, Maputo, Nairobi." The goal of this three-year initiative (2010–2012) was to improve watershed governance for climate change adaptation and enhance resilience and adaptive capacity of vulnerable and marginalized groups, especially women. It was supported by the Climate Change Adaptation in Africa (CCAA) program—a joint initiative of the International Development Research Centre (IDRC) and the United Kingdom's Department for International Development (DFID) and its methodology included collaboration between students, civil society organizations, and academics as well as community-based research and environmental education. Project partners based in universities and several CSOs in Kenya, Mozambique, and South Africa worked together to achieve the following objectives: 1) To characterize the institutional framework for urban water governance in the three cities, and explain how different actors within this framework cope with climate change and variability; 2) To identify and test viable alternatives for enhancing civil society's role towards adaptation to climate change and variability by vulnerable groups (e.g. by developing education, training, and awareness programs); and 3) To share widely the knowledge generated for potential adoption by other cities in Africa.

According to the Intergovernmental Panel on Climate Change (IPCC), "Africa is one of the most vulnerable continents to climate change and climate variability. This vulnerability is exacerbated by existing developmental challenges such as endemic poverty, limited access to capital, ecosystem degradation, and complex disasters and conflicts" (IPCC 2007). Income inequality in South Africa, Mozambique, and Kenya is among the largest in the world. In all three countries, equity struggles related to water are growing in social, political, and ecological significance; this is both a symptom and a cause of urban vulnerabilities related to climate change.

In Mozambique, climate change is causing coastal erosion, the destruction of mangroves and resulting threats to the locally important shrimp fishery, periodic flooding along scenic coastal roadways, saltwater intrusion, and wind erosion. In cities there is desertification in food-producing areas, flooding in coastal slum areas, degradation of water quality in wells, and potable water scarcity. The sea level of the Indian Ocean is rising with concomitant coastal management problems in Maputo municipality (UN Habitat 2010). Mozambique's national water law (1991) considers all water as state-owned, to be governed by the state for the benefit of the population, with water access for people, sustainable water supply, and stakeholder participation in water management as priorities. Four water basin committees have been established in Mozambique on the same general model as in Brazil.

Like Mozambique, South Africa is implementing watershed committees or "catchment management agencies" (CMAs) to decentralize decision-making and create a framework for integrating the needs of all stakeholders in water governance. But data on women's engagement with water governance are hard to find; the 2010 guidelines of the Southern African Development Community do not even mention women or gender as an important factor in watershed participation (SADC 2010).

The municipal government in Durban, South Africa has developed a local climate change adaptation strategy; Durban faces coastal inundation and storm surges related to sea level rise, hotter temperatures and heat waves, changed rainfall and storm patterns, slum flooding and reduced drinking water supplies due to climate change. Local policy initiatives rely for effectiveness on awareness and capacity regarding climate change risks and adaptive responses in civil society. Environmental education and confidence-building through capacity-raising are recognized as crucial needs in this process. For example, the Inkomati CMA, in a watershed north of Durban which includes Maputo, has initiated outreach programs targeting farmers, women, and youth. This type of action research is well developed in Durban, partly due to the work of the Centre for Civil Society at the University of KwaZulu-Natal and its partner CSOs.

In Nairobi, severe infrastructure needs are being exacerbated by water supply fluctuations and slum flooding related to climate change. Just as in Maputo and Durban, environmental awareness and education to promote more equitable governance processes are required. As noted by the Kenyan delegation to the 2007 UN conference on climate change in Nairobi, Kenya's adaptation priorities include education, good governance, human resources development and training, institutional capacity-building and management change, public finance improvement, and better national resources management.

Our Climate Change Adaptation in Africa (CCAA) project was implemented by the following community-based organizations: the Kilimanjaro Initiative (KI) and Kenya Debt Relief Network (KENDREN) in Nairobi; Women, Gender and Development (MuGeDe) and Justiça Ambiental (JA) in Maputo; and Umphilo waManzi (Water for Life) and the South Durban Community Environmental Alliance (SDCEA) in Durban. The University of Nairobi (Nairobi), Eduardo Mondlane University (Maputo), and the Centre for Civil Society at the University of KwaZulu-Natal (Durban) provided academic research coordination and student supervision for this project. The majority of CSO and academic participants in this project were women, and the community workshops, meetings, and organizing activities attracted and included far more women than men—probably due to women's social roles that require deep knowledge of water and the environment. Very few of the homes in the African slums where we worked have piped water or sewers; women carry water in buckets from community standpipes or wells for cooking, laundry, bathing, cleaning, and gardens. Flooding due to storm rainfall can overwhelm precarious, often informal sanitation systems, requiring cleanup by women, and care of the sick, children, and seniors are usually also women's responsibilities.

The project focused on low-income areas of each city, as these tend to be most severely affected by periodic flooding and other climate change impacts. The capacity-building aspects of the project included training and research sponsorship for students and faculty in the partner universities; support for community-based research, workshops in low-income communities and secondary schools given by CSO staff along with university students, curriculum and materials development, and skills development within the partner CSOs; training of environmental educators and organizers; contributions to the pool of experienced and qualified community workers in each country; strengthening of all the partner institutions' capabilities to carry out international projects; and contributions to the international literature and professional knowledge concerning water issues, environmental education techniques, and community organizing for improved civil society involvement in governance. The networks built extend from local and community-based linkages through regional and national-level policy groupings to international academic and policy networks on civil society, watershed management, and governance.

Following internships with local CSO partners in the project, university students participated in exchange trips to other countries and also worked with CSOs there on climate justice projects. In this way, the students and the organizations gained new perspectives on different ways of meeting local challenges and on the ecological and political contexts for their work.[5]

The political process of policy development and implementation depends on the interchange between civil society groups, researchers generating information on current realities, and government. This project attempted to challenge the conventional notion that only educational institutions "produce" knowledge. Understanding community needs, and what helps particular civil society groups to see and act to strengthen their role in democratic governance, is something in which community organizations and CSOs have eminent expertise.

One objective of this project was to demonstrate how partnerships between academics and non-academics can be very stimulating and effective. This type of partnership encourages and allows the partner CSOs to reflect on and analyze their activities and to document learning by bringing student researchers into the CSOs as collaborators/interns. The partnerships also encourage universities to be more pragmatic about teaching and research and to "field-test" approaches towards community organization, equity, and capacity-building. Students committed to the project's goals of building participatory engagement by local people in municipal water decision-making were given practical opportunities to develop their skills, as a way of advancing each city's climate change preparedness. This project emphasized the integration and meaningful participation of women in formal decision-making processes to increase their resilience and ability to cope with climate change.

Specific examples of how climate change responses combine well with gender-aware community organizing, all of which were explored through this project, included the following:

- The Kilimanjaro Initiative (KI), a youth-focused NGO, upgraded a sports field in Nairobi's Kibera slum, on the banks of the Nairobi River, which helps prevent housing from being flooded during extreme weather events. In addition, KI organized sports leagues including young women athletes, community forums on sustainable water management and environmental education, and community and river clean-ups. Young women's leadership was central to their organizing.
- In Durban, women activists from Umphilo waManzi and the South Durban Community Environmental Alliance (SDCEA) coordinated "learning journeys" where government officials visited low-income neighborhoods to hear about local women's experiences with flooding, sanitation, and other types of climate change stresses. This helped them to bring these views into policy discourse.
- Maputo university environmental education students worked with intermediate school youth on after-school activities related to climate change such as neighborhood and river clean-ups, clearing drainage canals, and documentary film screenings on environmental issues. Participation by parents and teachers was also welcome. The majority of participants were women.

In the words of Mozambican environmental education student Neima Adamo:

> *Para que a problemática do meio ambiente seja de facto um assunto comum entre a sociedade, é fundamental que a informação sobre a necessidade de preservar*

> *e conservar o meio ambiente seja disseminada por todos os actores da sociedade nos diversos níveis sociais, desta feita entende-se Educação Ambiental não-formal às acções e práticas educativas voltadas à sensibilização da colectividade sobre as questões ambientais e à sua organização e participação na defesa da qualidade do meio ambiente.* (In order for environmental issues to be a common subject in society, it's fundamental that information about the need to preserve and conserve the environment be known by all actors in society, at all social levels; in this way non-formal Environmental Education can be understood as all actions and educational practices aimed at sensitizing the community about environmental questions and organization and participation to protect environmental quality.)
>
> (Adamo 2012, 29)

Action to strengthen women's participation in water management

Both the Sister Watersheds project and the CCAA project, as well as the Green Change Project, explored practical methods of increasing women's "voice" in water management processes. Table 7.1 lists and summarizes some of the ways to do this. It has been our experience that these sorts of workshops, training programs, exercises, and strategies are adaptable and can be appropriate in a wide range of circumstances in both the Global South and the Global North.

As noted above, these types of grassroots water and climate change programming and organizing address long-standing, difficult gender inequities which have proven intractable despite decades of work, but which are nonetheless increasingly important in times of climate change. By creating spaces within local communities where women can share their knowledge of local water and climate conditions, develop confidence and respect for each other's abilities, and work together to devise and implement solutions, this sort of watershed-based organizing addresses gendered social inequities by making and calling for fundamental changes in governance. In feminist political ecology terms, this allows women to make use of their gendered environmental knowledge to work to advance their rights and responsibilities through networked grassroots political activism.

Conclusions

Engaging women and applying their expertise is fundamentally important for long-term climate change adaptation, particularly during environmental crises. Women's knowledge of local ecological and water conditions must be shared and utilized in local, national, and international decision-making processes—for reasons of both justice and efficiency. Democratic mediation of equity conflicts related to water, and sustainable long-term management of water resources, are only possible through civil society's participation in water governance. Moreover, because climate change mainly manifests itself through storms, floods, and

Table 7.1 Participatory and inclusive processes

Ways of bringing women into water management, starting at the community level
• Community mapping: residents collectively draw maps of important water features in the neighborhood and how extreme weather affects them; then discuss. • Photo-voice: community members photograph local scenes significant to them, in relation to climate change effects, and discuss/share with government officials. • Water dialogues: local residents discuss specific water issues with government officials, in forums facilitated by civil society organizations. • Water walks/storytelling parades: using waterways as a focus, local residents and visitors tour stream beds and floodplains, stopping for special presentations on local history, wildlife, food production, music, community assets, etc. • Collective storytelling workshops: community groups use selected "props" to spark creativity as they compose a joint story about the local watershed, sharing personal memories and expertise. • Community-based water monitoring: community groups work with government authorities to monitor pollution, biostatus, and flooding/drought in local waterways. • Water conflict mediation training: special CSO-run training programs address water conflicts (e.g. over standpipe access, fugitive emissions of pollutants, riparian rights, etc.) through skills training and information on how to access government supports. • Water harvesting: CSOs work with household members to develop ways of retaining rainfall from roofs and yards for home and garden use, and spread related practices. • Community gardening: CSOs locate space and train and organize community members to plant and water gardens for collective food production. • Leadership training: CSOs work with government and watershed committee officials to develop inclusivity training and confidence-building workshops for women, to welcome and facilitate their participation in water governance institutions.

droughts, climate change adaptation starts within watersheds and engages various scales of ecological and social interaction.

Community-based, "people-centered" approaches to climate change are crucial for adaptation strategies that address social and gender inequalities and allow women to serve as active agents of change in their communities. As Toulmin has argued:

> Past history shows that the poor and vulnerable do not get a fair share of resources unless they can mobilize effectively and there is parallel pressure on the powerful to make decisions in favor of the many, not the few. . . . This means making sure that the voices of ordinary (people)—women, men, young, old, farmers and slum dwellers—are heard loud and clear as the policies and institutions for addressing the most challenging of global problems are developed.
>
> (2010, 152)

Although community-based interventions offer unique opportunities for community involvement in climate change adaptation decision-making processes, and have the power to enhance the adaptive capacity of individuals, households, and communities, this approach is not perfect. Community projects tend to have short timescales, small budgets, and limited scope—i.e. they address some locally relevant climate change impacts, but not necessarily other contextual development challenges (Ludi et al. 2014). Thus, the projects mentioned in this chapter faced some important limitations.

The CCAA project, for example, ran for only three years, which is simply insufficient considering the time required to identify vulnerable communities, liaise with community leaders and elders, introduce the project to the community, identify one or more direct impacts of climate change through participatory community workshops, and begin implementing the project. Similarly, the project focused on a handful of *direct* climate change impacts—e.g. water scarcity, floods—but not on other drivers of vulnerability, such as poverty. More specifically, the project did not provide assets to community members or directly help to improve their livelihoods; instead, the project focused on raising awareness of climate change, educating community members on the climate change impacts facing their communities, improving community access to information, and facilitating women's interaction with other stakeholders, including government and private actors. The six-year Sister Watersheds project similarly focused on environmental education, confidence-building for political engagement, and local green community development. Whether these project activities were sufficient to substantially improve local people's capacity to adapt to or affect long-term changes (climatic or otherwise) is uncertain, and probably untestable. However, by strengthening local civil society organizations and reinforcing their water and climate justice related initiatives, these projects may have laid the groundwork for ongoing socio-political activism related to both climate change and deeper "development" issues.

Cross-cultural collaboration is another area of challenge and growth for projects such as ours, and for all who intend to address climate justice. Since our work has involved communication and collaboration across various kinds of difference—race, class, gender, the academic-grassroots divide, language, ethnicity, nationality—we have been aware of the need for sensitivity and attention to these differences throughout, and we have tried to create both formal and informal spaces for discussion and interpersonal sharing as a way of building trust and confidence in each other to facilitate our work together. For example, at the CCAA project's first meeting where most of the CSO and academic partners were meeting each other for the first time, we obtained special funding which allowed the African partners to visit Brazil together, make "field visits" to see successful public engagement and green community development initiatives there, and spend informal time together as well as more formal meetings where we discussed our joint goals for our project. We scheduled a session on gender at this first meeting, in order to share our understandings of gender and its importance in our project. Both the CCAA and Sister Watersheds projects involved student internships with CSOs, international student exchanges with CSOs in other countries, language instruction for exchange students, faculty field supervision of students' off-campus work, joint writing projects for students and junior faculty team members, and technical support visits to share organizing ideas from the Global South in other countries including Canada—all methods for bridging various kinds of difference and building team members' experience with how to do so. Our project started from the premise that local CSOs are already doing great work, so we should support them with funding and by building local and global university connections; by doing this we avoided many cross-cultural mistakes, because local organizations were in the lead.

The biggest cross-cultural challenge in any "development" project relates to money. As long as the budget comes from an international organization with its own objectives, conditionalities, and reporting requirements, and as long as the budget is administered through a university with its special bureaucracies and regulations, projects have a constant struggle to maintain good will, solidarity, and a semblance of participatory process. Over the course of our work, we encountered a number of frustrations related to funding delays and rules, which highlighted the importance of good language skills on the part of key project participants (the coordinator and staff responsible for finance and reporting in each partner organization), full transparency on all requirements and budget changes, and good groundwork to establish trust and a common sense of project goals and shared responsibilities. It is easier to bear these day-to-day hassles when all collaborators share the privileges and challenges of working together on a global problem like climate justice.

Existing development challenges—such as poverty, gender, and structural inequalities—and inadequate infrastructure aggravate the climate vulnerability of the poor, and of many women in particular. This chapter has summarized some ways that women are working together on climate education and water governance, helping to inspire and generate related strategies in other places which address both climate-related and underlying structural inequities.

These stories demonstrate the essence of feminist political ecology—what Ariel Salleh calls an "embodied materialist understanding (which) is indispensable to the trans-discipline of political ecology . . . the vital citizenship politics of ecological feminism" (2009, 6). Women start in their local communities, build on their gendered local knowledge, share skills and experiences, respect leaders as they emerge, work to build political rights for women through grassroots activism, and communicate with others elsewhere who are facing related and similar challenges to build movements at ever-larger scales within watersheds and other ecological and social structures.

As Salleh notes:

> an ecological feminist perspective emerges from praxis – action learning. . . . The global majority of women – being mothers and care givers – are culturally positioned as labour right at the point where humanity and nature interact. . . . It is certainly no exaggeration to say that the entire machinery of global capital rests on the material transactions of this reproductive labour force. . . . [T]hese agents of complexity are practicing both an alternative economics

> and an alternative epistemology. . . . The bearers of ecological and embodied debt are thus not simply victims of capitalist patriarchal institutions, they are leaders, and their people's science is one for the global North to emulate.
> (2009, 7–8)

Especially in times of climate change, this pattern of global organizing and leadership is urgent and compelling.

Notes

1 The authors acknowledge support from the Canadian International Development Agency through the Association of Universities and Colleges of Canada (Sister Watersheds Project in Brazil, 2002–2008) and the International Development Research Centre and U.K. Department for International Development (Climate Change Adaptation in Africa project, 2010–2012). This chapter is based on an article which appeared in the *Journal of Cleaner Production*, vol. 60, December 2013, pp. 188–194, "Women and Water Management in Times of Climate Change: Participatory and Inclusive Processes" by Patricia Figueiredo and Patricia E. Perkins, and is used with permission from Elsevier.
2 See Jane-Finch community website at www.jane-finch.com/about.htm. Accessed February 19, 2013.
3 See Green Change Project slide show presentation at www.google.ca/search?q=green+change+project&oq=green+change+project&aqs=chrome..69i57j0l2.2776j0j8&sourceid=chrome&espv=210&es_sm=91&ie=UTF-8. Accessed February 25, 2013.
4 Majora Carter is a South Bronx activist and international speaker who advocates and organizes for green urban renewal in low-income neighborhoods. She won a MacArthur "genius" grant in 2005. See www.ted.com/speakers/majora_carter.html, www.majoracartergroup.com/, http://yorkwestadvocate.wordpress.com/2011/02/22/environmental-activist-speaks-about-green-change-in-our-community/, and www.nytimes.com/2013/04/05/nyregion/a-hero-of-the-bronx-majora-carter-is-now-accused-of-betraying-it.html. Accessed November 5, 2013.
5 See the project's reports at http://ccaa.irisyorku.ca. Accessed February 25, 2013.

References

Adamo, N. 2012. *Relatório de estagio: Sistema de abastecimento de água e sameamento do meio: uma analise comparativa.* Universidade Eduardo Mondlane, Faculdade de Educação, Mozambique.

CIDA (Canadian International Development Agency). 2002. *Gender equality and climate change: Why consider gender equality when taking action on climate change?* Ottawa: Canadian International Development Agency. www.acdi-cida.gc.ca/INET/IMAGES.NSF/vLUImages/Climate%20change3/$file/Gender-2.pdf. Accessed July 11, 2013.

Elmhirst, R. 2011. Introducing new feminist political ecologies. *Geoforum* 42: 129–132.

Ensor, J. and R. Berger, eds. 2009. *Understanding climate change adaptation: Lessons from community-based approaches.* Bourton on Dunsmore, UK: Practical Action Publishing.

Faucheux, S. and C. Hue. 2001. From irreversibility to participation: Towards a participatory foresight for the governance of collective environmental risks. *Journal of Hazardous Materials* 86: 223–243.

Few, R., K. Brown, and E.L. Tompkins. 2007. Public participation and climate change adaptation: Avoiding the illusion of inclusion. *Climate Policy* 7(1): 46–59.

Gonçalves, B.C., B. Cavalcante, C. Gomes Pastor, A.L. Gomes, A.M. Pino Bustamante, C. Yamazaki Saravalle, F. Gonçalves Silva, J.H. Badi Zappala, L. Blaud Ciola, and S.M. Ribeiro. 2008. *Agroecologia urbana e práticas sustentáveis*. São Paulo: EPARREH. www.seaembu.org/docs/FANZINE_2008_Colorido_site.pdf. Accessed March 2, 2013.

Hinchcliffe, F., J. Thompson, P. Shah, J. Pretty, and I. Guijt, eds. 1999. *Fertile ground: The impacts of participatory watershed management*. London: Earthscan/IT Publications.

Holland, A. 1997. The foundations of environmental decision-making. *International Journal of Environment and Pollution* 7(4): 483–495.

IPCC (Intergovernmental Panel on Climate Change). 2007. *Climate change 2007: Synthesis report*. www.ipcc.ch. Accessed March 2, 2013.

JFCFC (Jane-Finch Community and Family Centre). 2012. Greening our communities through Live Green Toronto grants. www.toronto.ca/livegreen/pdf/jane_finch.pdf. Accessed March 2, 2013.

——. 2013. Green Change builds its dream home at 2999 Jane St. http://janefinchcentre.org/content/green-change-builds-its-dream-home-2999-jane-st. Accessed August 19, 2013.

Lorimer, E. 2011. The Green Change Project story. Unpublished paper, York University, Toronto.

Ludi, E., S. Wiggins, L. Jones, J. Lofthouse, and S. Levine. 2014. Adapting development: How wider development interventions can support adaptive capacity at the community level. In *Community-based adaptation to climate change: Scaling it up*, eds E.L.F. Schipper, J. Ayers, H. Reid, S. Huq, and A. Rahman, 36–74. London: Earthscan.

Mirza, M.M.Q. 2003. Climate change and extreme weather events: Can developing countries adapt? *Climate Policy* 3: 233–248.

Moraes, A. and Perkins, P.E. 2007. Women, equity, and participatory water management in Brazil. *International Feminist Journal of Politics* 9: 485–493.

Neumayer, E. and T. Plümper. 2007. The gendered nature of natural disasters: The impact of catastrophic events on the gender gap in life expectancy, 1981–2002. *Annals of the Association of American Geographers* 97: 551–566.

Paavola, J., W.N. Adger, and S. Huq. 2006. Multifaceted justice in adaptation to climate change. In *Fairness in adaptation to climate change*, eds W.N. Adger, J. Paavola, S. Huq, and M.J. Mace, 263–278. Cambridge, MA: MIT Press.

Perkins, P.E. 2003. Public participation in ecological valuation: How policies can help it happen. Paper presented at the conference of the Canadian Society for Ecological Economics (CANSEE), Jasper, Alberta, October 16–19.

——. 2004. Participation and watershed management: experiences from Brazil. Paper presented at the conference of the International Society for Ecological Economics (ISEE), Montreal, Canada, July 10–14.

Powell, R. 2010. *Vision: Centre for Green Change pathways to green jobs program*. www.nb.lung.ca/downloads/Rosemarie%20Powell.pdf. Accessed March 2, 2013.

Projeto Bacias Irmãs. 2007a. Agentes comunitarios de saude. Video showing community health groups and activities. www.youtube.com/watch?v=rqg7MMckLcU. Accessed January 9, 2012.

——. 2007b. Encontro estadual de educação ambiental.Video showing project activities at an environmental education conference. www.youtube.com/watch?v=9XzS3oLt-48. Accessed January 9, 2012.

——. 2008. *Balanço do projeto*. São Paulo: ECOAR. www.ecoar.org.br/web/files/files/Balanco_do_Projeto_Bacias_Irmas_VERSAOFINAL.pdf. Accessed January 9, 2012.

Rocheleau, D., B. Thomas-Slayter, and E. Wangari, eds. 1996. *Feminist political ecology: Global issues and local experiences*. London: Routledge.

SADC (Southern African Development Community). 2010. Guidelines for strengthening river basin organisations: Stakeholder participation. Gaborone, Botswana: SADC. www.limpoporak.com/_system/DMSStorage/3411en/SADC%20guideline_stakeholder.pdf. Accessed March 2, 2013.

Sagoff, M. 1998. Aggregation and deliberation in valuing environmental public goods: A look beyond contingent pricing. *Ecological Economics* 24: 213–230.

Salleh, A. 2009. *Eco-sufficiency and global justice: Women write political ecology*. London/New York/Melbourne: Pluto Press/Spinifex.

Sister Watersheds Project Final Report. 2008. *UPCD Tier 2 End-of-Project Narrative Report*, May 2008. www.yorku.ca/siswater. Accessed November 5, 2013.

Teixeira, D. de Lima, M. Ferraz Duarte, and P. Morimoto. 2008. *Manual de metodologias participativas para o desenvolvimento comunitário*. São Paulo: ECOAR. www.paulofreire.org/wpcontent/uploads/2012/CCP_Mat_Ref_Livros/manual_de_metodologias_participativas_para_o_desenvolvimento_comunitario_VERSC383OFINAL.pdf. Accessed January 12, 2012.

Terry, G. ed. 2009. *Climate change and gender justice*. Rugby, UK: Practical Action Publishing in association with Oxfam GB.

Todd, A. 2011. Climate change and water governance in the greater Toronto area. http://ccaa.irisyorku.ca. Accessed March 2, 2013.

Toulmin, C. 2010. *Climate change in Africa*. London: Zed Books.

UN Habitat (United Nations Human Settlements Program). 2010. *Climate change assessment for Maputo, Mozambique*. Cities and Climate Change Initiative. Nairobi, Kenya. www.unhabitat.org/pmss/getElectronicVersion.aspx?nr=2977&alt=1. Accessed July 11, 2013.

van den Hove, S. 2000. Participatory approaches to environmental policy-making: The European Commission climate policy process as a case study. *Ecological Economics* 33: 457–472.

Wilson, M.A. and R.B. Howarth. 2002. Discourse-based valuation of ecosystem services: Establishing fair outcomes through group deliberation. *Ecological Economics* 41: 431–443.

8 Women's contributions to climate change adaptation in Egypt's Mubarak Resettlement Scheme through cactus cultivation and adjusted irrigation

Dina Najjar

A feminist political ecology of climate change

This study examines the site-specific, social, political, ecological, and economic contexts that shape agricultural practices and policies in two desert resettlements of Sa'yda and Intilaq, also called the New Lands that form part of the massive Mubarak Resettlement Scheme (MRS) in Egypt. Within the political-ecological context of the New Lands, it takes a particular focus on practices and policies that enable women settlers to cope with climate change (Elmhirst 2011; Peet and Watts 2002; Rocheleau et al. 1996; Zimmerer and Bassett 2003). Environmental impacts often differ along gender lines within communities, and therefore, women must be directly involved in any attempt to remedy them (Agarwal 2000; Buechler 2009; Quagliiariello and Hamdy 2013; Sowers et al. 2011). This chapter contributes to the literature that examines gender in relation to environmental change and water resource struggles by offering a perspective on the Middle East.

The need for a gender analysis is heightened because marginalization along gender lines continues to be a major problem in the New Lands and requires adaptive measures, particularly in the face of climate change and the nascent Revolution. This region remains one of the most gender unequal areas in the world and is currently undergoing rapid social and political change (Haddad and Shideed 2013; Kaplan et al. 2011; Sowers et al. 2011; Waterbury 2013). The ongoing Egyptian Revolution highlights the degree of gender disparity including increased levels of violence against women. Considerations for gender inequalities in light of government policies, ecology, social upheaval, and climate change are therefore particularly important for poor women in the Middle East.

In the following sections, I situate the region and the settlement areas in terms of their climate and water characteristics. I then describe the particular characteristics of the two MRS settlements and compare women's strategies related to climate change adaptation. A discussion of the methods used follows identifying the value of anthropological methods for conducting research on climate adaptation. The policies that relate to agricultural practices and processes are analyzed with relation to how these policies impact women's adaptive capacities. To conclude, I discuss how this study contributes to climate adaptation literature in Egypt and beyond, including the policy implications as they relate to both the limitations and possibilities for strengthening women's active contributions to climate change adaptation in the MRS and more broadly.

Water and climate dynamics in the region and in the research sites

The Arab world is one of the regions most vulnerable to climate change. It is projected to have 10 to 30 percent decrease in precipitation in the coming years, leading to a subsequent decline in groundwater replenishment and seriously overexploited aquifers (Haddad and Shideed 2013; Sowers et al. 2011). In addition, increased temperatures and rising sea levels in coastal areas are expected to compound production problems in the influential agriculture sector. Fertile agricultural land is expected to either disappear into the Mediterranean Sea or become seriously affected by increased salinity levels (Haddad et al. 2011; Haddad and Shideed 2013). In Egypt alone, agriculture constituted 13.4 percent of GDP and employed 27.1 percent of the population in 2013 (CAMPAS 2013; Sowers 2011). The high population density in Egypt puts additional pressure on land and water resources.

Adaptation to climate change is often framed in terms of water use in the Middle East, where agriculture is the biggest consumer of water (Haddad et al. 2011; Waterbury 2013). The Middle East in general is the most water-stressed region in the world. More than half of its countries have average per capita water availability lower than the water scarcity threshold (Sowers et al. 2011). In Egypt, the Nile River is expected to experience decreased flow in the long term due to decreased rainfall in the watershed (Beyene et al. 2010). Increased temperatures in the Nile watershed are also leading to higher rates of evaporation and are likely to further exacerbate the effects of declining rainfall (Sowers et al. 2011). Furthermore, increased population pressure and water demands by upper riparian countries are expected to add to the water scarcity crisis in Egypt (Conway 2005). The Ministry of Water Resources and Irrigation (MWRI) predicts that by 2050, half of the Nile's water will be needed by Egypt's people to fulfill their potable water demands alone (Head of Horizontal Expansion, MWRI, 2013, personal communication).

Government responses to climate change in the region are, for the most part, focused on tapping into new sources of water, such as aquifers for groundwater pumping, and desalination of seawater (El-Sadek 2010; Zeitoun et al. 2012). In contrast, improvements to efficiency, such as drip irrigation, are only sometimes considered in water provision policies and are infrequently implemented. Sustainable and equitable distribution of water is a huge challenge due to strong lobby groups' abilities to maintain the status quo (Sowers et al. 2011; Zeitoun et al. 2012). These groups include large landowners, the political elites, and the ministries of irrigation that are largely influenced by the two aforementioned groups. The

Revolution that began in January 2011, however, has resulted in attempts by the marginalized to shake up these power imbalances by, for example, cutting off water access to all people in resettlement areas in protest and as part of demands for more equitable water distribution. Paradoxically, women living in resettlement villages often state that power imbalances lead to their exclusion both from government water programs and from revolutionary protests over water resources.

In addition to improving water management, other potential climate change adaptation strategies include integrating livestock and crop production, diversification, genetically improved material of seeds and livestock, and conservation agriculture (Haddad et al. 2011). These strategies, however, continue to be ineffectual as a result of inadequate policies, markets, research, and extension support (ibid.). These strategies also lack a focus on gender variations in adaptation capacities.

In Egypt, cultivating the desert, also called land reclamation, is a practice that is increasingly being employed for adaptation to climate change (Adriansen 2009). Desert land reclamation as promoted by the government entails tapping into aquifers found in the vast desert regions that make up 96 percent of the nation's land mass (Sowers 2011). The rising sea level of the Mediterranean basin is expected to lead to losses in homes and agricultural lands in the delta region. Up to 10 percent of the Egyptian population (about 6 million people) will be affected by seawater rise in the Mediterranean, and up to 15 percent of agricultural land in the delta could be lost to the sea. Government officials consider that desert land reclamation will compensate for this loss.

Although a growing body of literature is examining inequalities related to climate change impacts and implications for adaptation in Egypt and more generally the Middle East (see, for example, Malm and Esmailian 2013; Sowers et al. 2011; Zeitoun et al. 2012), very limited research has explored the gender dimensions of climate change impacts and adaptation strategies. A feminist political ecology perspective that addresses women's contributions to climate change adaptation can inform broader policy debates in Egypt. Accordingly, this chapter looks at the contributions of women to climate change adaptation in the MRS, specifically in light of government policies and the recent Revolution, and identifies ways to strengthen these adaptation strategies by comparing women's experiences with farming in two desert resettlements that differ markedly in socio-economic and biophysical contexts. Because political ecology research in general is often critiqued for a biased focus on political issues at the expense of the ecological dimensions (Walker 2005; Zimmerer and Bassett 2003), in this chapter I integrate both the ecological and political aspects of women's climate change adaptation practices.

The MRS, the settlements, and women settlers

Initiated in 1987, the MRS aimed to resettle millions of Egyptian people out of the overpopulated Nile valley and delta, or what is termed the 'Old Lands,' into 'New Lands' in the desert. The MRS project aims to alleviate poverty, to reduce unemployment and population pressure in the Nile valley and delta, and to increase food security and exports for the nation (Adriansen 2009; Meyer 1998) (see Figure 8.1).

Each settler in the MRS is provided with 5–6 acres of desert land, a shared water pump, and a homestead in a nearby village. The price of the property is determined by the category of settler. The New Lands are considered virgin lands with no previous inhabitants or land users (Adriansen 2009). These claims of

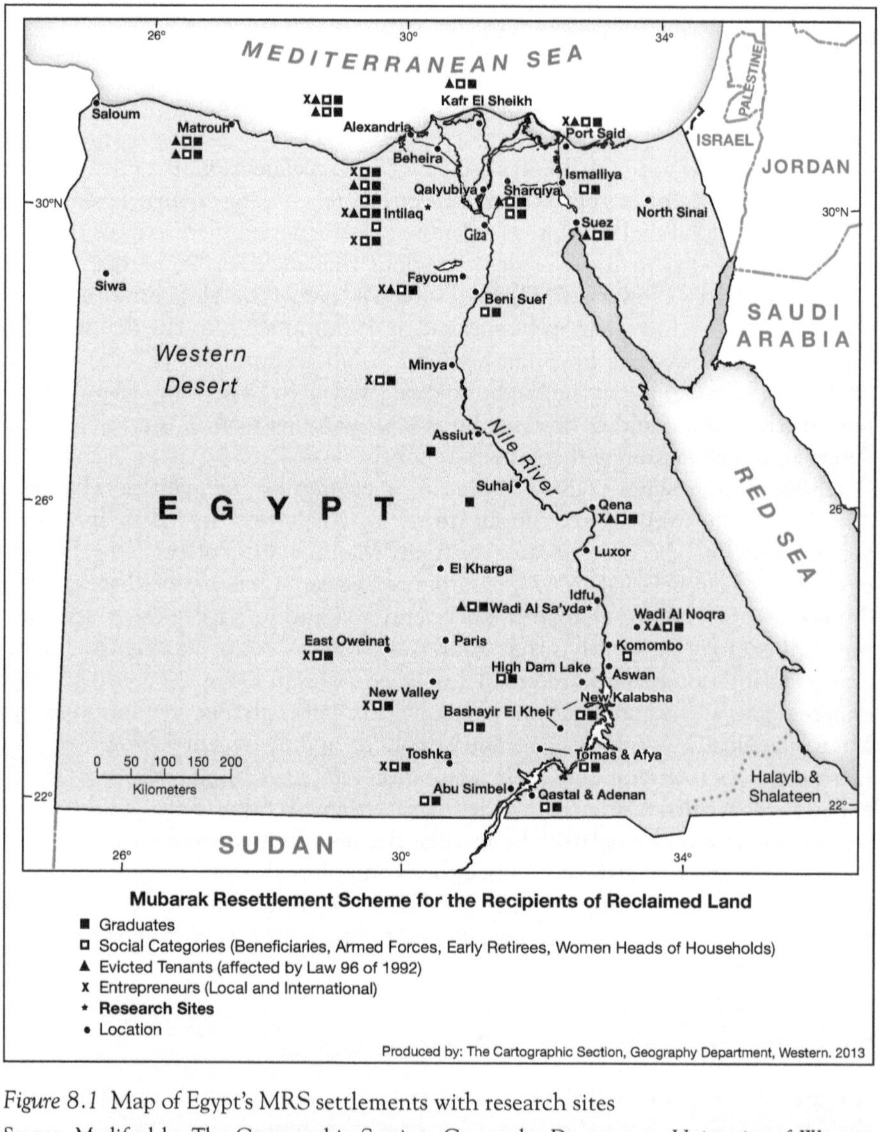

Figure 8.1 Map of Egypt's MRS settlements with research sites

Source: Modified by The Cartographic Section, Geography Department, University of Western Ontario from a map of the MRS provided in Arabic by Egypt's Land Reclamation Sector (2010)

virgin lands, however, are not verified, but are fueled by the belief that no vegetation[1] existed in these areas prior to desert farming.

Bringing land into production, or land reclamation, occurs in several stages. First, the land is leveled by the General Authority for Reconstruction Projects and Agricultural Reclamation (GARPAR) after which it is handed over to the Ministry of Agriculture and Land Reclamation (MALR). Leveling is essential in this process, because some lands are undulating, and the agriculture is dependent on irrigation. The MALR distributes land to the poor in collaboration with local governorates (political territories akin to states or provinces). Afterwards, settlers are instructed to leach their lands of salts thoroughly and then, with time, enrich their soils using organic material (such as animal manure and crop residues) as well as artificial fertilizers. This process requires about three years for the lands to properly convert to soils suitable for the desired crops. The main crops promoted by the government include wheat, fruit trees, and fodder. Fodder is especially encouraged because of its relation to raising animals that need constant care thereby persuading settlers to relocate permanently into the New Lands. Due to large costs for water and other inputs, it is not uncommon for settlers to operate at a financial loss in the early stages of reclamation. During these years, in many of the MRS settlements, the World Food Program (WFP) provides settlers with food aid until their lands start producing.

Research context and methods

This study draws on research conducted over a 14-month fieldwork period in 2010–2011 with follow-up research in 2012 and 2013 in two desert settlements: Sa'yda, a settlement in Upper Egypt, and Intilaq in Lower Egypt (see Figure 8.1). It uses ethnographic methods that offer valuable tools to uncover impacts of climate change and related adaptation strategies by means of thorough examination of local processes, behaviors, and relationships (Barnes et al. 2013). Identifying local innovations in adaptation to climate change, particularly those related to women, provides valuable lessons for up-scaling and for making visible the critical roles of women in arid regions. In its study of everyday practices and micro-politics within communities, feminist political ecology meshes well with the ethnographic focus on the local and particular (Truelove 2011). I used a qualitative, extended ethnographic approach in Cairo, Sa'yda, and Intilaq to study women's contributions to climate change adaptation in the resettlements. This included multiple data collection methods, long-term field research including observation (in daily activities and during on-farm visits), interviews, and focus group discussions, as well as information cross-checking over a 14-month residency in MRS communities (Bernard 2006; Marcus 1995). Multiple sources of data are important in qualitative research for validation purposes and to demonstrate different ways of knowing (Bernard 2006). Emerging themes from interviews and focus groups informed follow-up questions and discussions with participants at various levels (local and national officials, as well as local settlers).

Fieldwork in Sa'yda was mostly carried out in the 'Widow Village' of Samaha, a women-only village with 303 'widowed women.'[2] Women landholder informants in Intilaq were from three different villages. The two settlements differ in poverty rates, agricultural systems, services available, and types of women landholders (see Table 8.1).

A main difference between the communities in women's adaptation strategies is linked to their access to land. In Sa'yda, women accessed land mainly as Women Heads of Household Landholders (WHHLs), and in Intilaq, most women gained land access as Graduate Women Landholders (GWLs). Graduates are holders of a degree or diploma from a high school, technical school, or university.

Table 8.1 General characteristics of Sa'yda and Intilaq

Description criteria	*Intilaq in Lower Egypt*	*Sa'yda in Upper Egypt*
Age of settlement	Settled in 1999	Settled in 2002
Height above the Nile	17 m	46 m
Type of agriculture	Profit-oriented, fruit production, as well as peanuts, potatoes, and vegetable production	Subsistence-oriented (wheat, barley, and alfalfa) with cattle seen as symbol of wealth
Type of irrigation and source	Drip and sprinkler irrigation from Nile water and artesian wells	Flood irrigation from Nile water and/or drainage canals
Total area	750,000 acres	28,193 acres
Rainfall	2–3 times a year, about 15 mm a year	2–3 times a year, about 15 mm a year
Unemployment rates	8–12%	Higher than any other rural area in Egypt
Poverty[a]	15.3%	40.2%
Major development agencies involved	WFP and IFAD	WFP
Micro-credit	Available from the local agricultural bank, with an interest rate of 10%	Available only from the Local Agricultural Cooperatives (LACs), with an interest rate of 3–6%
Training	Gender-mixed and frequent	Mostly men and infrequent
Markets	Small-scale weekly (e.g. alfalfa (*berseem*), okra, cantaloupe) in nearby local markets Government mills and agricultural banks, which are more certain and reliable, thereby constituting the main market outlets	Export, local, and tourist-oriented

Note:

a Source of these poverty rates is the World Bank (2009), which is defined as whether the households or individuals have enough resources or abilities today to meet their needs.

Striving for as much diversity as possible in terms of age, religion, marital status, education, and ethnicity, a total of 55 women landholders (15 GWLs and 40 WHHLs) were selected for this study.

Sa'yda and Intilaq

In Sa'yda the MRS aims at alleviating poverty of WHHLs with young dependents by providing these women with a source of income, while in Intilaq the main goal of the MRS is to reduce unemployment for the graduate youth. This difference in resettlement policy is not surprising given that Upper Egypt (the location of Sa'yda) is home to the largest proportion of the poor in the country. Upper Egypt is home to 60 percent of those living in poverty and 80 percent of those living in severe poverty[3] (World Bank 2009). While women in Intilaq are educated, young, and predominantly came from wealthier governorates in Lower Egypt, women in Sa'yda are mostly illiterate, poor, are relatively older, and predominantly came from the poor Aswan Governorate in Upper Egypt.

Markets are another point of difference between the two settlements. Intilaq is well connected to tourism (such as restaurants in Cairo and Alexandria) and export markets. Many GWL settlers in Intilaq cultivate fruits for tourist restaurants and exports. Conversely, Sa'yda has poor access to markets, and most WHHL settlers focus on cultivating wheat mainly for subsistence with the crop residue used as fodder for animals. When there is an excess in wheat production, it is often sold to government mills. The Sa'yda settlement is thinly populated and, as a result, transportation of crops to markets (particularly vegetables, fruits, and alfalfa) is done on an individual basis and is expensive. Many settlers complain that transport costs offset possible profits. Wheat, however, is a 'strategic'[4] crop and can easily be sold to government mills and agricultural banks. Cattle and small ruminants are the main sources of income in Sa'yda. In fact, animals are *mal* meaning money in Sa'yda.

Other main differences between the two settlements were related to training opportunities for farmers and access to government loans. While training in desert farming methods was provided on a continuous basis in Intilaq, it was only provided sporadically in Sa'yda and almost exclusively to male settlers. This training is essential, since most farmers in the resettlements are accustomed to flood irrigation and farming on different kinds of soils. The limited training in desert farming has led to a situation where many men and even more women are unable to reclaim and farm their lands due to a lack of appropriate skills.

Intilaq settlers are provided loans on a reliable basis from the Agricultural Bank and the International Fund for Agricultural Development (IFAD). However, in Sa'yda, loans are mainly available through Local Agricultural Cooperatives (LACs) that provide only sporadic access to credit through a revolving fund. In Sa'yda loans are equivalent to only half the amount provided by the Agricultural Bank and IFAD in Intilaq. The revolving fund capital was provided by the WFP to the LACs based on a nominal fee collected from settlers for accessing food aid. In particular in Sa'yda, WHHLs are encouraged to apply for loans, but some sons-in-law take advantage of this preference for poor women and obtain the loans for their own purposes by using the names of their mothers-in-law.

Water management in the settlements

In each settlement, water is pumped up into desert areas using pumps that are maintained and run by the MWRI into a main canal, or *tar'a ra'isiya*. The Sa'yda settlement, for example, has a series of five water pumps that eventually elevate water 46 meters above the Nile into settlers' fields. These water pumps are maintained by the MWRI and work on a rotational basis, three pumps at a time. During the summer time, the high temperatures lead to malfunctioning in the pumps and water shortages are exacerbated. Each main canal branches into a sub-canal, or *tar'a far'iya*, to irrigate individual fields through smaller pumps distributed to the settlers by GARPAR. Each pump is shared by a group of four landholders. Some settlers are provided with electric pumps while others share a diesel pump. The maintenance of these on-farm pumps is the responsibility of the farmers.

Each settler is also provided by GARPAR with sprinklers, drip lines (only in Sa'yda), and irrigation rods (soil probes).[5] These irrigation technologies are intended to conserve water. Compared to the usual use of flood irrigation in the Old Lands, these technologies are expected to significantly conserve water to the extent that GARPAR believes that there is no need to build drainage canals. For farmers to irrigate using these technologies, a constant level of water is required in the canals for operating sprinklers, rods, and drip lines on a daily basis. While the maintenance of the main canals is the responsibility of the MWRI, maintenance of the sub-canals is the responsibility of the LACs.

Although women landholders, and settlers more generally, are provided with the same irrigation technologies, the irrigation methods in the two settlements differed markedly. WHHLs in Sa'yda frequently had their pumps stolen reportedly by young men and eventually were forced to switch to flood irrigation. The women-only status of the village led it to become susceptible to theft as thieves felt that these women were not able to defend themselves and could be easily robbed. As a result, almost all women landholders lost their electric pumps. The few whose pumps were not stolen often could not use their pumps anyway, because the electric transformers were stolen. "Even bricks and utensils get stolen when you leave your house for a few days. We sleep and wake up in fear in this village," explained Om Lateef (WHHL in Sa'yda).

Water governance

Participation in governance related to water access is mediated through two organizations located in each village: Water Councils and LACs. The LACs' budgets are derived from annual membership fees of landholders, and income-generating activities such as tree nurseries and the rental of machinery. As a condition for food aid provision, the WFP imposed quotas for women's participation in water governance during the initial stages of resettlement. On average, one woman per village joined the respective village LAC as one of the seven executive members. With time, women dropped out of the LACs because they felt that their participation as women was ineffective in male-dominated conversations, or they were not reelected, they got married and their husbands prohibited them from participating, and/or the engineers simply did not replace the women members who dropped out of the councils.

Since the WFP departure in 2011, Village Engineers, the organization that oversees the activities of the LACs, has not solicited women's participation on the LACs. In all of the Sa'yda settlement LACs in 2011, with the exception of the Widow Village, there was only one woman executive member. The situation for women's participation on LACs is better in Intilaq. Women in Intilaq, in contrast to those in Sa'yda, have more mobility and occupy farming roles in their areas. Even after the departure of the WFP, many LACs in Intilaq had up to two women executive members.

Water Councils, set up by the MWRI, also solicit women landholders as participants. Like the LACs, each Water Council has on average one woman member. The women-only village of Samaha in Sa'yda, however, has all female executive members on both committees. This village has 303 WHHLs who are divorced, widowed, abandoned, or married to husbands who are disabled. I will explore the extent and efficacy of women's participation on these committees, particularly related to women's abilities to obtain their water needs, in the section of the chapter addressing government policies.

The differences in services available to settlers, with Lower Egypt having far better access, are part of broader regional inequalities in government and development organizations' focus and spending within the settlements. Furthermore, the younger graduate settlers were regarded highly by the government as strategic to provide services and training to increase Egypt's exports and alleviate unemployment. The price of land sold to the graduates is therefore one-third of the price offered to people in other subsidized categories, including in poorer WHHLs.

In Sa'yda, as a result of all the hardships faced by both men and women for irrigation and farming, only 5 to 50 percent of settlers, depending on the proximity of the village to a main city, currently reside in the settlement. Their absence has harsh penalties, because settlers who do not immediately cultivate their lands or occupy their houses face eviction. During the early (and most unproductive) stages of land reallocation, however, most of the Sa'yda settlers resided in the settlement and, in fear of being evicted, farmed their lands regardless of productivity. In Intilaq, currently approximately 70 percent of landholders have relocated to the desert settlement, with many of the GWLs maintaining strong ties with their original hometowns to visit specialist doctors, buy clothing, and visit family members and loved ones.

All these differences in context between Sa'yda and Intilaq (e.g. government and development agencies' favoritism, markets, farming systems, and poverty levels) (also see Table 8.1) allow for a detailed understanding of circumstances that enable or impede women in their adaptation to climate change in this region.

Women's strategies to adapt to climate change

Despite the many political, social, and economic limitations women face in the settlements, many women are actively contributing to climate change adaptation by cultivating cactus crops that are more suited to arid soils, *Opuntia Ficus Indica F. Inermis* (see Figure 8.2), utilizing drip and sprinkler irrigation, and using alternative sources of irrigation.

Cactus, a women's crop: from the lazy farmer crop to a cash crop

As stated previously, farmers who did not cultivate their lands faced the threat of eviction. In Intilaq, the settlements initially had no schools, so many GWLs who were mothers of young children had to stay with their children in the Old Lands for their schooling. To avoid eviction, these women adopted prickly

Figure 8.2 Cactus plantation, *Opuntia Ficus Indica F. Inermis*, of a GWL in Intilaq settlement

Photo credit: François Molle, Institute of Research for Development and International Water Management Institute pear cactus crops because the cactus greens the land and requires minimal fertilization, weeding, and irrigation. GWLs either hired workers to cultivate their lands or went to the settlement for a short period of time every now and then to tend to their cactus orchards. Cactus in the Intilaq initially had the reputation of 'a lazy farmer's crop' with little economic value. Yet the GWLs did not care about this reputation, because they initially planted the cactus to avoid eviction and relocation into the hostile conditions of the New Lands.

Fruit production from cactus, many GWLs reported, has since changed their lives from being laborers on other people's farms to being able to provide their daughters with decent bridalware[6] for their early marriage and to elevate their status in the eyes of their in-laws.

The tourist industry (in Alexandria and Cairo) provides the main market for cactus fruit to flourish into a cash crop. Although the tourist market for prickly pear fruit has been hurt in the last few years by the political unrest, local demand remains steady. The cactus is also very suitable to the desert conditions and makes excellent use of scarce water and low fertility conditions of the New Lands. As such, this 'women's' or 'lazy farmer's' crop has the potential to relieve pressure on Egypt's exhausted surface and groundwater resources.

Drip and sprinkler irrigation: women-friendly irrigation methods

In addition to innovative desert crops, women have also adopted efficient irrigation methods for the desert settlements. Adoption of such methods came about largely due to social and physical limitations for women to use other types of irrigation. Most male farmers irrigate the land using small diesel pumps and surface irrigation, two methods that are reported to be physically prohibitive for women in both settlements. The former Head of Intilaq settlement, for example, views drip irrigation and electric pumps as one of the positive differences between the Old and New Lands with respect to women's roles: "due to sprinkler and drip irrigation and electric pumps, women can irrigate the land now as opposed to before when flood irrigation and small diesel pumps were used" (October 2010, personal communication). In Sa'yda, however, while the electric pumps made irrigation more accessible to women as the pumps operate with an on/off switch and do not require contact with the body, many women report theft of electric pumps. Without the electric pumps, they cannot irrigate their lands as easily by themselves. As Om Badawi, a WHHL in Sa'yda explained,

> When we had electric pumps, irrigating the land was very easy. You just press a button, and your land gets irrigated. Now, we need to hire someone to irrigate the land, and sometimes they lie: they would say that they irrigated the land when they really had not.
>
> (April 2011, personal communication)

There are also stories circulating about women being killed or severely hurt in the resettlement areas from operating small diesel pumps. Before resettlement,

Figure 8.3 GWLs in Intilaq settlement in field carrying tree for planting
Photo credit: Dina Najjar

in their hometowns, men had usually operated diesel pumps, so women had much less training and experience in using them. Furthermore, the long dress and long veil (*hijab*) that women wear (see Figure 8.3) could easily get caught in the turbine of the diesel pump reportedly choking one woman to death. Many women also perceive irrigation through diesel pumps to be too dangerous. Those women who operate the diesel pumps feel that their bodies were 'devalued' (Truelove 2011). For example, some complain that their backs or 'spines' are harmed as a result of walking on wet ground and often falling into the steep areas next to the pump; others are horrified when their *hijabs* get caught in the turbines of the diesel pumps since there is a possibility of getting severely injured.

Because of the many social and physical dangers of operating diesel pumps, women prefer to use electric pumps (despite dangers of theft) that are a much more efficient irrigation technology, contributing to more efficient use of water resources. Furthermore, the operating costs of diesel pumps are ten times higher than the operating costs of electric pumps. Electric pumps also have lower carbon emissions compared with the diesel pumps and reduce farmers' dependence on oil and diesel. Isolated areas in Egypt, such as the Sa'yda settlement, often face shortages in diesel and oil, which are needed to operate irrigation pumps, leading at times to crop failures.

Irrigation with inferior water

Finally, in addition to cactus crops and irrigation techniques, women also innovate by using drainage water for irrigation. All the MRS beneficiaries were provided with government land and housing located at the tail end of the water canals and farther from cities. Lands with better access to water at the head end of canals are sold to entrepreneurs, for prices set at 22 times higher than the prices at which land on subsidized terms is sold to the beneficiaries of the MRS (including the GWLs and WHHLs). Entrepreneurs in the context of Sa'yda and Intilaq are local Egyptians with higher incomes who are often working in the Gulf States or investors from the Gulf who employ local Egyptians for market-oriented agriculture.

In Sa'yda, the government, however, was unable to attract entrepreneurs to buy desert lands. These lands were often left vacant and were eventually taken over and farmed by relatively financially well-off individuals who were also called entrepreneurs and who were also primarily individuals who had worked in the Arab Gulf, mostly in Saudi Arabia. This type of entrepreneur, however, does not belong to a formal category. They did not obtain the land directly from the state. Because these entrepreneurs were informally occupying the land and farming it, the rules of farming in the Sa'yda settlement did not apply to them.

Sugarcane, bananas, and rice are forbidden crops in the New Lands because they consume large amounts of water, but informal entrepreneurs in Sa'yda cultivate sugarcane crops regardless of official regulations. In Intilaq , conversely, the entrepreneurs are a formal category, but the government turns a blind eye to their cultivation of bananas. The informal entrepreneurs of Sa'yda and the formal entrepreneurs of Intilaq consume most of the water allocated to the two settlements. Water scarcity at the end of the canals where the MRS beneficiaries are located was exacerbated with time due to an increase in the total area under cultivation by entrepreneurs.

As a result of these situations, many women landholders in Sa'yda irrigate their lands using drainage water and in Intilaq using artesian wells. Both water types are high in salts and, consequently, could lead to soil salinization over time. While GWLs in Intilaq supplement their irrigation with artesian well water with Nile River water, many WHHLs in Sa'yda had no access to Nile River water because they were provided with land at the end of the irrigation system and thus depend exclusively on drainage water. With adequate access to Nile water for leaching salts, both alternative irrigation sources contribute significantly to water conservation and efficiency, as well as sustainable agriculture in the settlements.

Related government policies, practices, and consequences for adaptation capacities

The government's practices sometimes overlapped with, and at other times conflicted with, women's adaptation strategies. The MRS aims to relocate millions of people out of the overpopulated Nile valley and delta. While the government wants settlers to relocate permanently to the New Lands, many GWLs and WHHLs were unable to relocate because of the lack of services, schooling for their children, and isolation from family networks that were important to women. Men who have less domestic and childcare responsibilities are able to relocate to the New Lands while their families stay back in the Old Lands. Many GWLs selected cactus as a crop that can be managed without relocation and with minimal financial and labor investments. In an indirect way, then, the government's settlement policies that more directly favor men's mobility also have facilitated the adoption of cactus as a crop for these women.

Although the policy of mandatory farming facilitates the adoption of cactus, other government policies are not supportive of this crop. The government does not encourage planting of cactus: "We do not focus on cactus. It is not a strategic crop. It is called 'the lazy farmer crop.' No research or extension has ever been done for cactus," explained the former Head of Intilaq (October 2010, personal communication).

Government policies related to cultivation of barley represent another example of ones that do not favor women's innovations. Despite the importance of barley as a subsistence crop throughout the Middle East, this crop is not subsidized in settlement areas (Ahmed et al. 2010). Women have also noted that barley could be a more suitable crop for the salt-tolerant desert conditions in the Sa'yda settlement where farmers grow mainly annual field crops and have very limited government support. Barley could replace the focus on wheat in Sa'yda where wheat yields are restricted by soil salinity and inadequate water supply. Women in Sa'yda, who are mostly elderly and poor, are limited in their abilities to provide labor and, subsequently, to leach out the salts adequately from their lands with sufficient irrigation water. These women in particular are interested in the ecological and financial benefits but have yet to be supported sufficiently to grow salt-tolerant crops such as barley.

In Egypt there is a lack of extension services or research on cactus, barley, or any other drought-tolerant crop, allowing faulty beliefs about cactus cultivation to prevail. Male farmers and government officials believe that intercropping with cactus (as opposed to fruit, where intercropping is widely practiced with fava beans, peanuts, alfalfa, and wheat crops) will lead to bitter crops. In other arid regions with low fertility, legume intercropping has proven to be particularly beneficial (Bezner Kerr et al. 2007). However, there is no research that proves that alley cropping with cactus leads to bitter crops (see Alary et al. 2007). Furthermore, women in the settlements have found cactus to be beneficial socially, economically, and ecologically. It allows women to maintain their social networks and domestic responsibilities (i.e. they can still live in the Old Lands and bring children to school), while also requiring low water usage and minimal crop maintenance, and women have found new markets for the cactus fruits.

In relation to irrigation, government programs also limited women's adaptive capacities to desert farming. In Sa'yda, while the government provided women with electric pumps for sprinkler and drip irrigation systems, the theft of the electric pumps often limited the ability of women to irrigate their lands in this way. Putting women all together in one village, which the WFP thought would be good for delivering women-exclusive services and enhancing their participation in public life, led to their increased vulnerability to theft after which many women were discouraged from farming their lands.

Another government-related sphere that affects women's adaptive capacities is in the water-related decision-making committees, particularly the LACs and Water Councils. In both settlements, many women executive members have difficulty asserting their water demands reflecting both gendered inequalities and that the committees' roles altogether are ineffective. In Sa'yda, for example, Water Councils are nominally formed, but have no concrete roles and consequently were largely ineffectual. When asked about who the members are, farmers in general in Sa'yda had no clue as to who their representatives were on government Water Councils.

In the LACs, women in Intilaq do not have leverage due to their small numbers (usually only one woman at most on each committee). As Om Fathi, a GWL and LAC executive member explained,

> I am alone on the committee. I proposed many things, like having a school bus and mobilizing for building a secondary school, but no one listens to me. Perhaps, if there was another woman member with me, I would be listened to.
>
> (August 2011, personal communication)

The demands made by LACs for irrigation water were also not answered by the government, limiting their effectiveness and forcing many women farmers into despair. Upon realizing her crops would fail because of her inability to pay for sufficient irrigation water, Om Rabee' (a GWL and Water Council member) stated, "The government sells water to the entrepreneurs, God is my savior and in him I confide" (October 2010, personal communication). Om Rabee' was struggling to repay her loan because of inadequate water for irrigating her land that she had planted, following government advice, with grapes and orange trees—crops that require significant water for irrigation. These findings are congruent with Sowers et al.'s (2011) and Zeitoun et al.'s (2012) findings on newly discovered water sources in the MENA (Middle East and North Africa) region benefiting elite groups exclusively. In my study on women this was the case with the entrepreneurs.

Although all the executive members of the LAC in the Samaha village are women, their participation remained nominal (Agarwal 2001, 2010). The Village Engineer on many occasions reported that he "does not want to trouble" these executive members with meetings and responsibilities, and, instead, he has the president sign blank checks at his disposal. Indeed, every time I meet with the executive members they complain that coming to the LAC meetings is a hassle, and very few of them know what their roles as executive members entail. Furthermore, the Head of the Sa'yda settlement rarely meets with the women-exclusive Samaha village LAC, but often meets with the male-dominated LACs very late at night making it difficult for any women members on the respective LACs to attend. It is dangerous for women to travel to meetings in the isolated areas of the settlements. It is also not socially appropriate for women to meet late at night with men, and most women have many cooking and other domestic responsibilities in the evening hours.

Finally, since the onset of the Revolution, women have also been excluded from the informal and protest strategies used by poor male farmers to assert their water needs. For example, in July 2011 men blocked irrigation water passage to everyone in the Sa'yda settlement as protest over inadequate water supply causing problems for farmers during the peak irrigation season for wheat crops. In fear of ouster, the Head of Sa'yda and Governor of Aswan were obliged to meet male farmers' needs. Cutting the water supply put enormous pressure on the state (including the LRS (Land Reclamation Sector), Governorate, and the MWRI) to provide settlers with irrigation water, potable water, and reduced water pumping costs on the LACs. Women in Sa'yda, however, did not participate in blocking the passage of irrigation water. As they have not seen as many direct benefits, women have been less optimistic about the Revolution and many do not feel that protest is a safe or appropriate domain for women. The limited participation of women in formal protest is most notably felt in the women-only Samaha village. The whole village was excluded from the benefits of the protests in Sa'yda, and, through 2014, Samaha still suffered from a lack of potable water.

Conclusions

In examining women landholders' adaptive capacities to climate change in Sa'yda and Intilaq on the biophysical, cultural, political, and institutional fronts, it became clear that women are largely disadvantaged by local norms (especially in Upper Egypt settlements) and government policies. Since women's participation on LACs and Water Councils is only nominal for the most part, and due to the limited effectiveness of these organizations more generally, they are not able to claim adequate access to Nile water.[7] Late night meetings of executive members and a lack of appreciation for women's roles in public life have further excluded women and have limited women executive members in their ability to assert their needs in the public domain.

In many cases, women settlers expressed that they have no political leverage in local institutions because of their low numbers on Water Councils and LACs, while in other cases, even those serving on all-women committees feel that their participation is burdensome and futile. Along the same lines, women did not participate in Revolutionary protests believing their participation to be ineffective. Despite women's many adaptive strategies for farming in the arid resettlement lands, their lack of political and public voice limits their ability to harness positive change. More research is needed to understand how women from different social classes and educational backgrounds can participate effectively in public life in order to access and assert control over natural resources in the Middle East. Although Agarwal (2001, 2010) asserts that in Asia, women-only cooperatives are effective routes for women's participation in natural resources management, the case of women-only resettlements in Egypt points to contradictory findings.

The location of women's lands at the end of water lines necessitated innovation in irrigation sources, which included inferior quality water, often salty, from irrigation drainage canals or artesian wells, thereby enhancing the recycling of Nile water. Supplementary irrigation from the Nile River, however, is essential in cases where inferior water sources are used to avoid a build-up of salinity in the soil surface to counterproductive levels and to ensure in general greater sustainability of the settlement. This study demonstrates the benefits of balancing the ecological and the political in political ecology studies (Walker 2005). Addressing unequal access to water as well as realizing an integrated, efficient use of drainage and Nile water is essential for achieving sustainable farming in both settlements.

There is a need to focus on drought-tolerant crops rather than the current focus on wheat for achieving resilient farming in Egypt. Alley cropping, better management practices, added value to cactus such as the use of cactus as cattle feed and in cosmetics create solid opportunities for improved livelihoods, integrated livestock and crop production, and a better match between the ecological conditions and cropping patterns (Alary et al. 2007). Introducing cactus for livestock feed could be successful for both government and settlers since the ability to raise livestock sways people to stay in the New Lands. This is a win-win strategy. A focus on barley would help the cropping system become more adaptive to higher salinity and water scarcity, especially in the Sa'yda settlement. Along these same lines, there is a need to make some alterations in diets to substitute for virtual water imports, particularly wheat imports (Yang et al. 2003). Cactus, unheard of in Upper Egypt, for example, could play a large role in stimulating a change in diet. Exchange visits by both men and women between Upper and Lower Egypt could help break gender-related taboos (since Lower Egypt is more gender progressive) and help in the adoption of cactus and the stimulation of increased consumer demand for cactus since water in Sa'yda is scarce, especially in villages at the end of water lines.

On the bright side, women in gender-mixed villages of Intilaq where water pumps were not stolen are enabled by the government to irrigate their lands through modern irrigation techniques over the long term: electric water pumps connected to women-friendly sprinkler and drip irrigation networks. Women in Intilaq and Sa'yda, prior to the theft of their pumps, had been able to adopt the new role of farm managers, partially due to the ease of irrigating by 'clicking the pump button.' In Intilaq, the electric pumps were still in use in 2014, but in Sa'yda after a brief period of using the electric pumps, they were stolen and women were forced to either irrigate with diesel pumps or hire an irrigator. Since women could not afford to purchase new pumps and the government did not replace them, many women in Samaha stopped farming their lands.

Processes of engaging societal actors or tapping into social capital to formulate adaptive strategies to climate change are key to water equality, efficiency, and conservation (Kumnerdpet and Sinclair 2011). The current literature on such topics is frequently gender blind, and even more so when addressing such issues in the Middle East (for exceptions see Skutsch 2002). The Egyptian Revolution provides the political urgency to focus on social capital for adaptive strategies and to address equity issues in water distribution, since crisis demands immediate action (Waterbury 2013). Government failure to recognize the adaptive capacity of women and to support their innovations, however, requires special attention. Women settlers and farmers have felt helpless in addressing water resource issues, even with the onset of the Revolution. Through a focus on the interconnected nature of gender, politics, and natural resources, this chapter contributes to feminist environmental analyses by addressing the gendered limitations to social institutions and natural resource decision-making, and by illustrating women's adaptation strategies to make the best use of social and natural resources under conditions of scarcity. Clearly, there is a need in this region for more societal support of women for their own adaptive strategies, such as promoting research on cactus production and increased security measures for women settlers engaging in modern (electric pump) irrigation methods.

Finally, this study opens up new grounds to re-conceptualize adaptation strategies to climate change in the Middle East in ways that move away from an exclusive focus on water use efficiency and tapping into new water resources towards examining inequalities in water access, women's inclusion, and alternative farming practices that are suitable to the agro-ecological conditions of dry areas (Haddad et al. 2011; Waterbury 2013). A feminist political ecology framework sheds light on the interplay between government planning, local innovations, inequalities, and environmental conditions in adapting to climate change; an understanding of these linkages and their consequences is necessary for finding solutions that are both sustainable and relevant to local contexts (Elmhirst 2011; Truelove 2011). The study demonstrates not only the dangers and inequalities felt by women farmers in this area, but also highlights the potential that women's innovations have to positively influence social and ecological sustainability in this region during times of political unrest and climate change. This work encourages policymakers and scholars alike to be more supportive of strategies that integrate environmental, political, and cultural solutions into climate change adaptation, and to support measures that tap into women's innovative yet often underutilized capacities.

Notes

1 In Toshka, for example, a reclamation project in the Western Desert (see Figure 8.1), there was no growth before land reclamation because the soil is very saline and rainfall is almost nil (Issam Bashour, 2014, personal communication). It rains about once every ten years in Toshka.

2 The term widowed woman is a collective term that could refer to a woman who is divorced, abandoned, widowed, married to a man with physical challenges, or with dependent parents.

3 Severe poverty is a condition characterized by a deprivation of basic human needs, such as food, health care, and education.

4 Wheat constitutes more than 40 percent of the diets of the predominantly poor population in Egypt. The government spends a good part of its budget paying for imported wheat, bought in volatile market conditions. After the food crisis of 2007–2008 and the toppling of the Mubarak regime, self-sufficiency in wheat production came back to the policy table, and wheat prices, at which state mills and banks buy from farmers, were set at higher values from previous years. The head of the LDU (Local Development Unit; government body part of the Ministry of Agriculture and Land Reclamation) in Sa'yda encouraged farmers to cultivate wheat on a continuous basis.

5 Sprinklers and rods are to be used for irrigating field crops (wheat, peanuts, alfalfa). Drip irrigation is to be used for irrigating tree seedlings and eventually orchards.

6 Bridalware is similar to trousseau or dowry. The bride's parents provide her with clothing for years to come, kitchenware, cupboard, washing machine, television, and carpets, as well as, sometimes, gold. As opposed to dowry, the bridalware belongs to the bride and is not a gift to the groom's family but is a contribution to the couple's new home.

7 Note that the Nile water in general is reported to be recycled many times on its way to the Mediterranean Sea, and through the process, the solute content increases reducing the quality of water with time (Barnes 2012).

References

Adriansen, H.K. 2009. Land reclamation in Egypt: A study of life in the new lands. *Geoforum* 40: 664–674.

Agarwal, B. 2000. Conceptualising environmental collective action: Why gender matters. *Cambridge Journal of Economics* 24(3): 283–310.

———. 2001. Participatory exclusions, community forestry, and gender: An analysis for South Asia and a conceptual framework. *World Development* 29(10): 1623–1648.

———. 2010. Rethinking agricultural production collectivities. *Economic and Political Weekly* 45(9): 64–78.

Ahmed, M.A., K. Shideed, and A. Mazid. 2010. Returns to policy-oriented agricultural research: The case of barley fertilization in Syria. *World Development* 38(10): 1462–1472.

Alary, V., A. Nefzaoui, and M.B. Jemaa. 2007. Promoting the adoption of natural resource management technology in arid and semi-arid areas: Modelling the impact of spineless cactus in alley cropping in Central Tunisia. *Agricultural Systems* 94(2): 573–585.

Barnes, J. 2012. Mixing waters: The reuse of agricultural drainage water in Egypt. *Geoforum* 57(November): 181–191.

Barnes, J., M. Dove, M. Lahsen, A. Mathews, P. McElwee, R. McIntosh, and K. Yager. 2013. Contribution of anthropology to the study of climate change. *Nature Climate Change* 3(6): 541–544.

Bernard, H.R. 2006. *Research methods in anthropology: Qualitative and quantitative approaches*. Oxford: Altamira Press.

Beyene, T., D.P. Lettenmaier, and P. Kabat. 2010. Hydrologic impacts of climate change on the Nile River Basin: Implications of the 2007 IPCC scenarios. *Climatic Change* 100(3–4): 433–461.

Bezner Kerr, R., S. Snapp, L. Shumba, and R. Msachi. 2007. Participatory research on legume diversification with Malawian smallholder farmers for improved human nutrition and soil fertility. *Experimental Agriculture* 43(4): 437–453.

Buechler, S. 2009. Gender, water, and climate change in Sonora, Mexico: Implications for policies and programmes on agricultural income-generation. *Gender and Development* 17(1): 51–66.

CAMPAS (Central Agency for Public Mobilization and Statistics). 2013. *Annual Book of Census Data for the Arabic Republic of Egypt 2013*. www.capmas.gov.eg. Accessed April 24, 2014.

Conway, D. 2005. From headwater tributaries to international river: Observing and adapting to climate variability and change in the Nile basin. *Global Environmental Change* 15(2): 99–114.

Elmhirst, R. 2011. Introducing new feminist political ecologies. *Geoforum* 42(2): 129–132.

El-Sadek, A. 2010. Water desalination: An imperative measure for water security in Egypt. *Desalination* 250(3): 876–884.

Haddad, N. and K. Shideed. 2013. Mainstreaming adaptation to climate change into the development agenda. In *Climate change and food security in West Asia and North Africa*, eds M.V.K. Sivakumar, R. Lal, R. Selvaraju, and R. Hamdan, 301–315. Amsterdam: Springer.

Haddad, N., M. Duwayri, T. Oweis, Z. Bishaw, B. Rischkowsky, A.A. Hassan, and S. Grando. 2011. The potential of small-scale rainfed agriculture to strengthen food security in Arab countries. *Food Security* 3(1): 163–173.

Kaplan, R.L., M. Khawaja, and N. Linos. 2011. Husband's control and sexual coercion within marriage: Findings from a population-based survey in Egypt. *Violence against Women* 17(11): 1465–1479.

Kumnerdpet, W. and A.J. Sinclair. 2011. Implementing participatory irrigation management in Thailand. *Water Policy* 13(2): 265–286.

Malm, A. and S. Esmailian. 2013. Ways in and out of vulnerability to climate change: Abandoning the Mubarak Project in the northern Nile Delta, Egypt. *Antipode* 45(2): 474–492.

Marcus, G.E. 1995. Ethnography in/of the world system: The emergence of multi-sited ethnography. *Annual Review of Anthropology* 24: 95–117.

Meyer, G. 1998. Economic changes in the newly reclaimed land: From state farming to smallholding and private agricultural enterprises. In *Directions of change in rural Egypt*, eds S. Hopkins and K. Westergaard, 334–356. Cairo: American University of Cairo Press.

Peet, P.R. and M. Watts, M., eds. 2002. *Liberation ecologies: Environment, development and social movements*. London: Routledge.

Quagliiariello, R. and A. Hamdy. 2013. Integrating gender into climate change response. International Conference: Policies for Water and Food Security in the Dry Areas, Cairo, Egypt, June 24–26.

Rocheleau, D., B. Thomas-Slayter, and E. Wangari, eds. 1996. *Feminist political ecology: Global issues and local experience*. London: Routledge.

Skutsch, M.M. 2002. Protocols, treaties, and action: The "climate change process" viewed through gender spectacles. *Gender & Development* 10(2): 30–39.

Sowers, J. 2011. Remapping the nation, critiquing the state. In *Environmental imaginaries of the Middle East and North Africa*, eds D.K. Davis and E. Burke, 158–191. Athens: Ohio University Press.

Sowers, J., A. Vengosh, and E. Weinthal. 2011. Climate change, water resources, and the politics of adaptation in the Middle East and North Africa. *Climatic Change* 104(3–4): 599–627.

Truelove, Y. 2011. (Re-)conceptualizing water inequality in Delhi, India through a feminist political ecology framework. *Geoforum* 42(2): 143–152.

Walker, P.A. 2005. Political ecology: Where is the ecology? *Progress in Human Geography* 29(1): 73–82.

Waterbury, J. 2013. The political economy of climate change in the Arab Region. Arab Human Development Report Research Paper Series. United Nations Development Programme. www.arab-hdr.org/publications/other/ahdrps/AHDR%20ENG%20Waterbury%20v3.pdf. Accessed April 24, 2014.

World Bank. 2009. Arab Republic of Egypt, Upper Egypt: Pathways to shared growth. Social and Economic Development Group Middle East and North Africa Region. Washington, DC: World Bank.

Yang, H., P. Reichert, K.C. Abbaspour, and A.J. Zehnder. 2003. A water resources threshold and its implications for food security. *Environmental Science and Technology* 37(14): 3048–3054.

Zeitoun, M., T. Allan, N. Al Aulaqi, A. Jabarin, and H. Laamrani. 2012. Water demand management in Yemen and Jordan: Addressing power and interests. *The Geographical Journal* 178(1): 54–66.

Zimmerer, K.S. and T.J. Bassett, eds. 2003. *Political ecology: An integrative approach to geography and environment-development studies*. New York: Guilford Press.

Part III

Stories, narratives, and knowledge production of socio-environmental change

9 Shoes in the seaweed and bottles on the beach

Global garbage and women's oral histories of socio-environmental change in coastal Yucatán

Anne-Marie Hanson

One morning in February 2011, I set out with the women of the composting group Las Costeras to pick up seaweed that washed ashore following "*nortes*" or the heavy northern winds that bring rains and stormy weather to the coastal communities of northwestern Yucatán, Mexico. While raking up the seaweed to be composted, one woman held up a shoe, laughed out loud and yelled, "I found one!" The rest of the women also laughed and told me that it was a competition to see who could find the first shoe of the day, since they generally encountered several shoes in the washed up seaweed, as well as many other discarded objects, quite often with foreign labels, nearly all plastic. They continued to laugh about all the random discarded things they found each time they cleaned up the seaweed, as another woman said jokingly, "well, we never find *pairs* of shoes, so we cannot imagine—where in the world are all these people walking around with just one shoe?"

I begin with this story, because it expresses two important points that will be addressed in this chapter. First, the story is one of many that demonstrate accounts of lived experience that help us to make sense of the world around us and to produce actions that confront the global garbage crisis as experienced by coastal communities. Such stories are especially relevant for understanding socio-environmental change in low-lying tropical coastal areas, where uneven development is increasingly extreme in both social and geographical terms (Castree et al. 2010; Smith 1984). Low-lying coastal areas disproportionately experience the majority of urban disasters, are at great risk for flooding due to sea level rise and intensified tropical storms, and also predominate among environmental conservation hot spots worldwide (Church et al. 2004; McGranahan et al. 2007: 18).

Second, the story of women laughing and joking while cleaning up seaweed and other people's garbage further points to the strange position garbage plays in the context of small cities on the northwestern coast of Yucatán. It highlights the extremely important, but highly underpaid and under-recognized, work of women in the mitigation of coastal hazards produced by overflows of both local and globally produced waste products. Many large-scale conservation interventions focus on nature protection in tropical coastal areas of the global South in order to conserve endangered species and seemingly exotic, out-of-the way landscapes such as those in Yucatán. These interventions often reshape localized places and subjectivities, and in Yucatán, tend to favor men's fishing and eco-tourism activities for coastal conservation and sustainable development. However, these conservation interventions usually don't address global political-economic structures that simultaneously encourage increasing plastics consumption and urbanization—two processes that allow old shoes and plastic bottle fragments to reach faraway places and disrupt local ecologies. In turn, coastal conservation policies focused on tourism and resource extraction activities tend to simplify or ignore local gendered power dynamics as related to "urban" household divisions of labor or gender norms regarding disposal of household, community, or globally produced garbage products.

Through sharing stories of two women's waste management groups and their relationship to activism and socio-ecological change in small coastal cities, I argue that women's stories and narrative oral histories are critical forms of knowledge for confronting garbage-induced water, health, and development challenges in tropical coastal areas. To address these issues, I begin with a broad discussion of recent dialogues in feminist political ecology that contribute to research on garbage and global marine litter and their effects on coastal livelihoods, health, and water resources. I then position garbage within the gendered and spatialized context of nature protection and urbanization in coastal Yucatán. I draw heavily from oral histories collected in 2010–2013 of women's experience in the home, in conservation space, and as participants in plastics recycling, seaweed composting, and wetlands cleanup programs. Their histories include what motivates them to participate in garbage work. The oral histories highlight social and physical challenges women faced while dealing with these waste management issues. I conclude by addressing the ways in which women's collective organizing through recycling and composting projects position them as local professionals who positively influence urban sustainability in coastal areas where both urban issues and women's work in conservation have long been excluded from policy and practice. I argue that women's oral histories of socio-environmental change express their embodied knowledge and the lived experiences navigating and confronting global patterns of consumption and conservation. The work done by women—as demonstrated through their histories—in turn directly influences regional waste management strategies by expanding grassroots cleanup organizations in politically underrepresented coastal areas. Their work further demonstrates the potential international significance of women's collective environmental organizing to confront the increasingly severe garbage-induced hazards in vulnerable coastal areas.

Feminist political ecology: linking global garbage, urbanization, and coastal hazards

Garbage as a socio-environmental issue is generally placed within larger discussions of sanitation in urban spaces, and it is increasingly recognized as a key development challenge throughout the global South (Medina 2010). Women and poor communities often take on the responsibility of waste reduction and recycling due to disorganized or absent state support for urban waste issues (Beall 2006; Hanson forthcoming). The lack of support for waste services or waste workers in many politically marginalized areas has also led diverse grassroots women and other community groups to organize together in order to seek environmental justice, political representation, economic opportunity, and better living conditions (Gutberlet 2008; Moore 2008; Rosado 2009). While there is a sizable amount of scholarship on women and waste, past research predominantly focused on women waste pickers and recyclers in very large urban centers (Beall 1997; Fergutz et al. 2011; Fredericks 2009; Huysman 1994; Mehra et al. 1996), and little research in this area has used a feminist political ecology framework to analyze the scalar connections of global and local consumption patterns, or focused on waste management practices in coastal areas.

Feminist political ecology links feminist methodological and theoretical perspectives with multi-scalar analyses of ecological, economic, and political power relations (Carney 1993; Jarosz 2011; Truelove 2011). Feminist political ecology also pays particular attention to the mutual embeddedness of hierarchies of power based on gender, class, ethnicity, and race in the politics of control and access to environmental resources, or the rights and responsibilities with relation to environmental pollution or decline, ranging from intimate to global scales (Elmhirst 2011). Building from earlier feminist political ecology contributions that focus on gendered analyses of political and ecological networks of power within neoliberal environments (Harris 2006; Nightingale 2011; Sultana 2009), this chapter draws from women's oral histories to critique the ways that women are excluded from coastal conservation-as-development processes that focus on specific gender-based activities in which not all can participate. This study has broader implications for the role of narrative oral histories in addressing local responses to the globally connected issues of increasing consumerism and rapid environmental change. In turn, the research also demonstrates how women's knowledge and practices shape globalizing processes of coastal conservation through grassroots management of urban waste issues that are also excluded from policy.

While most studies on waste focus on large (and often inland) urban centers, almost half of the world's population now live within 150 kilometers of a coastline, and by 2025 that figure is likely to double to over 6 billion (UN-OCEANS 2013). Furthermore, at the same time that modern urban citizens expect their places of work, play, and residence to be clean, ordered, and free of trash, the economic globalization processes that create sanitary urban spaces also produce an excess of garbage through expanding consumerism (Moore 2009). People who live in increasingly urbanized coastal areas experience the cumulative effects of environmental stress and health concerns from resource-intense activities in coastal and marine areas, inland urban development (Creel 2003: 5), as well as increased consumerism leading to further garbage[1] accumulation in estuaries, marine areas, and coastal wetlands. Decline and deterioration of coastal and marine resources are compounded by increasing volumes of non-organic solid wastes (i.e. garbage) and lack of controlled final disposal in coastal areas (Hanson forthcoming). Most locally produced solid wastes include various plastics from food and other packaging, sanitary wastes, discarded composite wastes from the local fishing industry (including plastics, metals, and other mixed materials), and other waste building materials from coastal hotels and houses.

Internationally, marine litter is one of the most pervasive pollution problems faced by the Earth today (UNEP-CAR/RCU 2008). Over 7 million tons of debris are estimated to reach the oceans each year, made entirely or partially of plastic and including bags, cups, packaging and containers, clothing, shoes, and sanitary and sewage-related plastic items (Fanshawe and Everand 2002; UNEP 2005). As much as 80 percent of plastic ocean debris comes from inland sources in the global North, such as under-secured dumps/landfills, riverine transport, untreated sewage and stormwater discharge, industrial and manufacturing facilities, and tourism (UNEP 2005; UNEP-CAR/RCU 2008). The other 20 percent comes from sea-based sources including cruise ships, fishing boats, oil/gas platforms, and shipping, military and research vessels (NOAA 2014). Furthermore, many discarded synthetic materials are extremely buoyant and can travel long distances, impacting environments, wildlife, and human health far from their origin (Mcllgorm et al. 2011).

In the Gulf of Mexico, a large amount of plastic waste washes ashore along with the floating *sargazo* (Sargassum seaweed) that grows in the mid-Atlantic Ocean Gyre. The seaweed and entrapped plastic wastes follow ocean currents and drift into the Gulf of Mexico. Recently, higher fluctuations in cold and warm temperature extremes on the ocean surface have caused longer growing periods for floating seaweed (meaning more quantity and larger plants), followed by an increase in frequency and intensity of tropical storms that push the *sargazo* onto the beaches. While some of this seaweed actually helps reduce beach erosion by holding the sand in place, communities located across the Gulf of Mexico have long removed rotting seaweed for tourism aesthetics (both visual and olfactory) and health purposes (i.e. rotting seaweed attracts many flies and other pest animals) (Gower and King 2011; Morton et al. 2004). The increased amounts of seaweed also bring more plastic waste from all around the world that is mixed together in the mid-Atlantic Ocean. In this sense, while global garbage may be one of the ugliest reminders that the world is indeed increasingly well-connected, it remains true that no matter where you live, most people live intensely local lives (Pratt and Hanson 1994).

Garbage and nature tourism in coastal Yucatán

Coastal Yucatán includes several small cities that share characteristics with many coastal cities worldwide and that currently confront formidable challenges to public and environmental health and to sustainable development in the face of a rapidly changing climate (Figure 9.1). These include increased in-migration

Figure 9.1 Map of coastal Yucatán

Map credit: Shipeng Sun, University of Illinois Springfield

and higher plastics consumption levels, coupled with increased coastal hazards, and the insufficient collection and inappropriate disposal of municipal solid waste (Hazra and Goel 2009; Medina 2005; Myers 2005; UN-HABITAT 2008). On average in Mexican cities only around 40 percent of collected waste is disposed of in secure landfills (Medina 2005; SEMARNAT 2009). Furthermore, waste management in small coastal cities of the Yucatán Peninsula represents the least organized waste collection in the region, and garbage poses growing problems for governments and citizens (Cob Chay and Ucán Chan 2010; SEDUMA 2009).

The garbage problem in coastal Yucatán is severe, and is a result of the gradual merging of social, political, and environmental networks that historically were—and for the most part continue to be—categorized and managed in separate urban or rural spheres. For example, economic and population growth in Mexico has caused increased urbanization and a modification in consumption patterns within communities of all sizes. While most coastal cities in Yucatán are still very small by global urban standards, the coastal population quadrupled in size in 20 years (1970s–1990s) due to state programs that encouraged inland farmers to relocate to the coast to participate in the once-growing fishing industry (Fraga 2004). Although local consumption habits are similar to those in large cities, most of this growth occurred without funds or infrastructure for urban services like drainage, sewer systems, or garbage collection. There is also no place for safe garbage disposal in the coastal wetlands; the land is not ecologically suitable for final disposal or containment. Most existing municipal dumps in coastal cities are very small (only able to contain 10,000 tons of materials) (DOF 2004, 2011), have few environmental or health safeguards, and have reached capacity or are currently overflowing with no additional stable land to build new ones. The growing quantity of locally produced garbage in the dumps is in addition to the various plastics and other solid wastes abandoned by Caribbean cruise ships or left behind by beach tourists. As a result, garbage in the dumps is burned in large quantities on a regular basis (Interview with Celestún dump manager, March 2011).

In addition to an increased coastal population, in the last 30 years, the state strongly promoted urbanized lifestyles and consumerism by linking these to concepts of a shared national and progressive identity via education and the media, which led to changes in consumer habits (Leatherman and Goodman 2005). For example, while many Yucatán residents in rural and semi-urban areas do not have drainage or garbage collection, the majority do have cable television and internet (INEGI 2010). Furthermore, individual consumption of plastic products has been heavily promoted throughout Mexico (and globally). With the exception of very remote rural areas, the few existing studies on consumption and garbage in Mexico reflect strong similarities in rural and urban consumption habits and reveal that around half of the solid wastes produced and collected are plastics (Buenrostro Delgado and Israde 2003). A large volume of household waste includes plastic drinking containers. As safe drinking water is not regulated in most areas, several national and international beverage companies provide bottled water, as well as other heavily advertised soft drinks and carbonated beverages. There has also been a dramatic increase in the consumption of "*comida chatarra*" or junk foods in plastic wrappers (e.g. chips, candies, and cookies) (Leatherman and Goodman 2005).

Intense development of the coastal tourism industry has meant increased availability of food products in plastic containers, leading to more garbage production by both residents and tourists. Residents of Celestún, Sisal, and Dzilam de Bravo note that there are very few garbage bins in town, irregular or no garbage collection in residential areas outside main tourist centers, and the beaches only have one or two public garbage bins, "except during peak tourist times when there are more" (Interviews, March, April, and September 2011). Much of the plastic waste finds its way to the ocean, to natural spring-fed estuaries, or to *cenotes* (freshwater sinkholes), later surfacing on coastlines and becoming entrapped in coastal wetlands. Garbage is a chief contaminant of freshwater sources and aquifers (Cob Chay 2011). Seventy percent or more of *cenotes* and *ojos de agua* (natural freshwater springs) are contaminated with garbage from solid wastes, agricultural runoff, and sewage (Notimex 2013).

In some small coastal cities, waste plastics and other scrap materials have been used as a source of landfill to stabilize homes located on the edges of wetlands, or to expand yards that slowly deteriorate due to their location on the swamps and in areas of increased annual flooding. Diagnostic reports for coastal tourism and conservation programs similarly address irregular urbanization and tourism development on the wetlands as the primary impediments to mangrove conservation in natural protected areas (García de Fuentes et al. 2011; Gold Bouchot et al. 2011). As one Yucatán delegate of Profepa[2] stated,

> Tires, plastic containers, and aluminum cans are commonly used to fill in mangroves and wetlands in the coast of Yucatán. Garbage and contamination are destroying the places where irregular settlements are built, but that is not the main problem. If you have mangroves with garbage the environmental impacts affect the entire coastal area – garbage permeates everywhere. Mangroves are converted into garbage dumps.
>
> (Sepulveda 2013)

These social, political, and material issues surrounding coastal garbage are further compounded by cumulative effects of global environmental change. Increasingly severe seasonal weather patterns affect livelihoods dependent on coastal resources. Gradual rise in sea levels (approximately 0.13 inches per year from 1993 to 2009) (Church et al. 2004) produces more extreme and larger areas of annual flooding, and intensified tropical storms erode beaches and wash ashore even more of the estimated annual 50,000 tons of marine litter (UNEP 2006).

While these indicators alone are alarming, they also contradict global images of coastal Yucatán as an eco-tourist paradise enveloped by internationally protected corridors of mangrove forests, marine reefs and coastal dunes home to large numbers of sea turtles, whale sharks, and tens of thousands of pink flamingos. Since the 1980s, the beaches and coastal wetlands of northwest Yucatán have been primary foci of many national and international conservation organizations. Particularly in the last 15 years, environmental protection programs funded by the World Bank Global Environment Fund and the Mesoamerican Biological Corridor have been supported by private (including Ducks Unlimited de México; The Nature Conservancy; Rare International; and Pronatura) and federal conservation programs (García de Fuentes et al. 2011). While sustainable development efforts in large urban areas of Mexico concentrate on issues such as municipal waste management, sustainable development in small towns and rural coastal areas is often instead focused on community-based resource management near national parks and within biosphere reserves (Fraga 2006; Garcia-Frapolli et al. 2009; Haenn 2005).

In the last 15 years, state, federal, and international conservation areas expanded and currently encompass or are adjacent to nearly all coastal Yucatán towns and cities (SEMARNAT 2008). Furthermore, there is an increasing institutional push by state, federal, and international environmental agencies to replace fishing—the main economic activity on the coast—with beach tourism and eco-tourism in state and national protected conservation areas (Capurro et al. 2002; Córdoba Azcárate 2006; Fraga 2009; García de Fuentes et al. 2011; Gold Bouchot et al. 2011; SEMARNAT 2008).

Along all Yucatán Peninsula coasts, overfishing, unsustainable tourism development, and rapidly increased migration to coastal towns and cities have degraded and eroded much of the beach and disrupted the coastal wetlands ecosystem. Due to the total elimination of mangroves in Cancun for mass tourism development and intense urbanization of the eastern coast of the Yucatán Peninsula, federal policies address environmental decline in northwestern Yucatán State through large-scale and top-down programs that promote conservation activities and establish and maintain state parks and federal biosphere reserves (García de Fuentes et al. 2011). Millions of dollars are raised by international conservation organizations to maintain these areas for research, restoration, and recreation. As money flows in to support conservation efforts in these natural protected areas, the region now receives millions of tourists each year. These tourists often arrive at the small coastal cities to visit beaches or take wetlands tours, bringing with them food containers and disposable items they do not take back home with them. Garbage production in small coastal cities increases four to ten times during peak tourist seasons (Interview with Celestún dump manager, May 2011). Particularly in high tourist seasons (March–April and December), these small coastal towns can have over 100,000 tourists daily who produce about 50 tons of garbage per day (Lliteras 2009).

The same conservation regulations that encourage "eco-tourism" to support coastal conservation also prohibit or highly regulate all forms of local resource extraction and urban development. Nearly all of the economic activities prevalent in coastal Yucatán are seasonal and heavily dependent on natural resources and government subsidies for support. Since the 1980s, the main subsidized activity in addition to tourism or eco-tourism is export fishing. This includes several types of export fish, such as octopus, sea cucumber, and other highly protected, but valuable commercial species. Although a large majority of coastal residents rely on government subsidies for their livelihoods, they are also predominantly directed at men's work. Many women participate in fishing activities not by catching fish, but by cleaning and processing fish. However, women are often locally perceived as housewives or helpers to men's activities (Radel 2012), and do not regularly receive government or private fishing or ecotourism-based employment subsidies (Hanson forthcoming). Additionally, since a main focus of national and state policies is on establishing and maintaining conservation areas for research and recreational use, the urban issue of garbage management was largely under-prioritized at all levels of government until recently. Very few of the small towns in coastal Yucatán had municipally funded or privately organized garbage management prior to the year 2000 (Interviews with residents and CONANP employees, November 2010, January–May 2011).

Solid waste is now recognized internationally as a threat to expanding conservation territories (Vásquez Avila et al. 2011), but the concern over garbage is not new to coastal residents. Women and local communities not only see the effects of garbage on landscape and community, but also directly experience the lived consequences of garbage decomposition in wetlands, such as soil degradation, and infectious diseases like cholera and dengue. While men spend much of their day fishing and outside the home, most women in coastal cities spend their time working in and around homes built on or near the wetlands (Hanson forthcoming). For these women, the abundant garbage on the streets, decomposing in wetlands, and entering groundwater supplies poses an additional source of vulnerability for the health and safety of themselves and their families. It was largely health concerns stemming from increased garbage and lack of municipal waste management that inspired women to organize.

Methods

This chapter is based on ethnographic fieldwork conducted in Yucatán between September 2010 and January 2013. The research included a variety of qualitative methods to inquire about garbage management and its relationship to urban development and environmental protection on the coast. Through interviews, oral histories, and participant observation of women's garbage activities, I learned the various ways that individual families managed their own garbage and participated in conservation or urban development activities. In this chapter, I focus on personal narratives and oral histories that highlight the motivations, struggles, and everyday experiences of women who are actively involved in community waste management through composting and recycling activities. In carrying out oral history interviews, I followed the concept of ecological oral history (NAU 2005), which allows an analysis of landscape change from the perspective of people who have used that land over time and which is embedded in people's life experiences (Robertson and McGee 2003).

I conducted 14 oral histories with coastal women to focus on their analyses of changes in the urban, social, and ecological landscape. The histories were collected in interviews that lasted over one hour each and were conducted by me in Spanish. I asked women to discuss what motivated them to begin and continue participating in community waste management, what experiences shape those motivations, and how their work in garbage cleanup has changed their livelihoods or altered the social and environmental landscape in coastal Yucatán. Narrative analysis was carried out with the understanding that stories and exact words of individuals can provide useful insights into social and spatial processes and events (Wiles et al. 2005). This type of analysis helps to produce information about socio-environmental change that is embodied and rooted in place by framing the interviews as the narrators' stories (Endres 2011), thereby expanding the possibilities to influence policy using culturally relevant and gendered knowledge from vulnerable areas.

Furthermore, instead of focusing on women's narratives as a way to reveal subjugated knowledge and silenced voices (Kesby 2005), these histories shift the focus to underscore how women in Yucatán became local professionals and environmental stewards who maintain clean spaces, and to put at the forefront women's motivations and stories of socio-environmental change (Cahill 2007; Jarosz 2011; Jenkins 2008). Using these women's exact words helps to situate their knowledge as embodied (Mohanty 2003), while also demonstrating the ways in which women's knowledge and daily practices are critical to long-term environmental sustainability in coastal Yucatán and have international significance in policy debates over the management of vulnerable coastal areas. By sharing the spoken narratives of women and situating these within broader socio-ecological and political contexts, my objectives are to address how individual women speak about their personal experiences as part of recycling and composting groups; to highlight how everyday experiences relate to state-society relationships and changes in conservation and city space; and to present the ways in which women reposition themselves as urban sustainability professionals (Hanson forthcoming).

Global garbage and women's stories of collective action

Between 1997 and 2002, in the towns of Celestún and Dzilam de Bravo, two small groups of coastal women organized to collect plastic bottles, compost seaweed, and organize wetlands and beach cleanup projects. Recycling, composting, and wetlands cleanup are all tasks that require complicated organizing and very hard work with little financial gain. Due to local gendered expectations where women previously did not often work long hours outside the home and because of general lack of support for garbage cleanup in small coastal cities, at times the women were highly stigmatized for working with garbage. People belittled them and called them "pepenadoras" or scavengers, and in some cases they were subjected to domestic violence and community resistance. For example, women recyclers explained how "men would go around and throw our collected plastics into the yards, wetlands, or streets next to our homes" (Interviews, June 2011). On several occasions, their recycling collection center was broken into and ransacked for usable building materials (Interviews, October 2011). In Dzilam de Bravo, women composters explained that there were at first many "delicate husbands" who resisted their work. They would burn the women's collected piles of seaweed before they were able to remove plastic wastes and prepare the seaweed for compost (Interviews, March 2011).

Las Costeras began as a group of 25 women who organized in 1997 in Dzilam de Bravo, motivated by the need to keep their town healthy, their beaches clean, and to attract tourists. It was the first group of its kind to collect the washed up seaweed in order to turn it into compost that is then later sold to local inland farmers or used in community gardens. The women also organized a small garden for group members on a plot of land near their composting facility located about 500 km inland from the port city (Interviews, April 2011). This group was the first composting group in Yucatán, and since has stretched to include over 400 women in similar organizations along the coast. As one woman told me: "Look, men here are only interested in fishing or tourism. We (women) wanted to do this, because it is embarrassing. People come here to see the beaches, not seaweed, and definitely not shoes!"

Although they do not earn large sums of money, the group has received temporary grants from the state environmental agency (Secretary of Urban Development and Environment, SEDUMA) and research institutions (the Center for Investigation of the National Polytechnic Institute, CINVESTAV). Each woman is paid something for doing her part in the process, be it collecting, drying, or preparing the compost. Women earn small sums of money, obtain compost to sell, and receive vegetables for their families grown in the community garden (Interviews, June 2011). Still, many women explained to me that they were continuously defending themselves and justifying why they would bother cleaning up the seaweed or other people's trash when they were barely paid anything. As one woman who was active in Las Costeras told me:

> You know, they used to tease us and say "they are dirty" and "like to play with garbage." . . . But the thing is, nobody really recycled before, or did anything about the trash on the beach. And we were blamed for it. [To clean up the beach] we used to bury the seaweed under the sand, but we can't bury the garbage. We thank God for all the tourists, because we need their business. But the tourists and the storms bring us trash, too. Shoes and bottles from all over the world, I mean, this stuff has writing on it in different languages I don't understand. And then, we keep hearing all this talk about sustainable development and conserving the environment, but those aren't my shoes, and no one I know even drinks Pepsi . . . , and it was just us cleaning them up!
> (Interviews, February 2011)

Each year Las Costeras and similar composting groups remove over 80,000 tons of washed up seaweed and 2,000 tons of garbage from the beaches along the coast. Women in some nearby towns joined this project after learning about it through state (SEDUMA) employees who visit towns near natural protected areas. In other towns, women learned of the project through family members and friends in Dzilam de Bravo (Interviews, June 2012). In March 2013 the group was officially recognized by SEDUMA and has a contract of ongoing support to supply compost to inland farmers for reduction of water use for irrigation. As the leader of the group expressed:

> This new work is a big commitment, but I know I have the support of all the group members. This is for us to stay healthy, to sustain our families, and to help our husbands when there is no fish . . . and to keep the beaches clean.
> (Interviews, January 2013)

Furthermore, soils in Yucatán are primarily derived from karst limestone and are very permeable. There are no rivers in Yucatán, making groundwater scarce and limiting the few crops able to grow near the coast or further inland (Vazquez-Dominguez and Arita 2010). Women and men in coastal towns note that with the increased use of compost, soils are enriched, and many more vegetables (especially carrots, peppers, and potatoes) are grown by local farmers and the women's group that are later sold in markets in coastal towns and in specialty markets in Mérida (Interviews, October 2012). As one member of Las Costeras noted:

> We have seen a big change. Before, we did not have much paid work and the community did not recognize us. But now, the agencies that used to just give money to fishermen and eco-tourism groups are noticing that we women also have important work to do. We clean the beaches, and our compost helps protect the environment for farmers. We can grow food that we couldn't before. While my kids might not like it, now we eat more carrots!
> (Interviews, October 2012)

In addition to the seaweed-composting group Las Costeras, another grassroots women's group similarly emerged out of women's concern for health and a clean community. In 2002, the recycling group Chen Kole 'Lob, meaning "only women" in Yucatec Maya, began with 17 women who collect plastic bottles and organize wetlands cleanup projects. The women organized together following an Oportunidades lecture in 2002 given by Dr Eduardo Batllori, a lead social scientist from CINVESTAV (Unidad Merida) and current director of the Yucatán SEDUMA. Several thousands of women in coastal Yucatán receive educational subsidies through the Oportunidades national development program, which gives monthly stipends to mothers living below the poverty line who have children enrolled in elementary or secondary school (SEDESOL and Oportunidades 2013). To receive the subsidy, the women are required to attend lectures every two months on nutrition, education, and public health. This particular talk focused on the relationship between garbage, coastal environments, and health risks. Women were especially concerned with the overflow of garbage in providing breeding grounds for dengue-carrying mosquitos and the links between waste decomposition and the cholera outbreaks that affected many coastal residents in the 1990s and early 2000s (Interviews, April 2011). The lecture also provided a meeting place to organize around the issue.

The women began their work by trading useable household items (e.g. toilet paper, rice, sugar, etc.) donated by local businesses and hotels for plastic bottles. Then, a local NGO donated tricycles and the women went around collecting bottles. Currently, in Celestún, almost everyone collects bottles and the women pass by every two weeks to collect them. Hardly any bottles go to the dump or are burned or buried anymore. In 2002, the women successfully secured a temporary work grant through the National Commission for Natural Protected Areas (CONANP). For all of the women interviewed, recycling is an activity that is not just a way of making a living, but also a way to keep the city clean and the community healthy. The women only earn 25 cents per kg of plastic bottles sold to the recycling plant in Mérida (the capital city of Yucatán, located 60 km inland). The women pay residents 20 cents per kg, so their net earnings are only 5 cents per kg (Interviews, August 2011). However, most members explained their motivations in working with the cooperative were not for monetary gain, but stemmed from previous employment and health concerns related to garbage:

> We wanted to invite the community to become interested in recycling, in garbage, so that it doesn't pile up. I used to work in a hotel, and when people arrive here they say that they saw pictures on the Internet and that Celestún was pretty and clean, but they arrive and see the town is dirty. So many visitors have said that. Too many. It is embarrassing.
>
> (Interviews, April 2011)

> I think as women we were always cleaning anyway . . . if we don't clean our yards every day, the winds or the water bring the garbage back into our homes. . . . It was terrible. Because, well . . . here in Celestún, garbage has brought us many problems . . . we have had cholera. We have had dengue.

Figure 9.2 Garbage in wetlands behind homes in Celestún

Photo credit: Anne-Marie Hanson

> Every day we saw people getting sick and even dying. The clinic was full and there was no room. There was too much garbage. If picking up the garbage can stop that, I want to participate.
>
> (Interviews, March 2011)

The women themselves view their work as building a better community in Celestún by promoting clean spaces, conscious behaviors, and healthy environments at the grassroots. Many women recruited their adult and teenage daughters into the recycling group, and several community members collect bottles, which now very rarely end up in lagoons behind homes. As one woman told me:

> At this point, currently, we cannot stop collecting recyclables, because if we do, it will all end up back in the lagoons. Today, in 2011, we are in this to make a better Celestún. And we also try to support those people who are not in our group. Right now in our community, there are people who don't have a way to get by, and now they collect bottles and make 50 or 200 pesos every two weeks. In all of this, we have made things better in the whole community. In 2003, none of this existed. . . . There is still garbage on the streets, but nothing compared to ten years ago, even five years ago. You don't see bottles on the streets anymore. And we cleaned up the beaches, too. . . . Ask anyone, ask the *municipio*, ask the park service, ask the businesses. NO one sees as much garbage around. We did that. . . . We planted the seed for Celestún.
>
> (Interviews, March 2011)

Chen Kole 'Lob has also come to be seen by their community as a group of local professionals who promote recycling and maintain clean spaces. The women have cleaned up over 10,000 tons of garbage from the wetlands, and recycled at least 2,000 tons of plastic containers every two weeks from yards, wetlands, and streets in Celestún. Chen Kole 'Lob also inspired similar projects in other coastal towns (for example, the towns of Sisal, Chabihau, and Chixchilub). Although women are not given all the credit for organizing better forms of waste management, CONANP, SEDUMA, and active NGOs credit them as important actors in current integrated waste management systems for the state of Yucatán (Vásquez Avila et al. 2011).

A local soda bottling plant and recycling center (BEPENSA) has begun to support more women's recycling groups in coastal towns by providing them with the materials and space for recycling collection centers, with the ultimate goal to provide women's groups with recycling centers in over 60 towns across the entire Yucatán Peninsula coastline (Pacheco 2010). Chen Kole 'Lob and similar women's groups now educate coastal residents about the health risks posed by plastic wastes and how to recycle. Residents note that with the cleaner streets, fewer people have fallen ill to dengue or other garbage-related diseases (Interviews, September 2012). As a result of these efforts, federal agencies (CONANP and SEMARNAT) have also begun to educate tourists on beaches and in natural protected areas about the damaging effects of tourist-produced garbage in local communities and natural areas through social media campaigns (see http://dsiapps.semarnat.gob.mx/#sinbasura).

Global garbage, feminist political ecology, and coastal women's stories of activism

As John Scanlan argues, "garbage is everywhere but, curiously, is mostly overlooked in what we take to be valuable from our lived experiences, and crucially, in the ways we organize the world" (2005: 9). As the urban population and tourism industry grow in coastal Yucatán, plastics consumption has increased without land or local government infrastructure to dispose of waste materials in controlled sites. As a result, "small cities on the coast often face the cumulative burden of decline and deterioration of coastal resources, while also experiencing infectious diseases and urban infrastructure problems directly related to overflow of garbage in these areas" (Hanson forthcoming).

In focusing on the ways that women's organizations deal with the overlooked problem of too much trash on the coast and in coastal waterways, this chapter puts at the forefront women's motivations for collective organizing in a way that positions them as catalysts for positive socio-environmental change. Rather than position women's words as subaltern testimonials that further normalize patriarchal and neoliberal state projects (Curiel 2007), women's histories reject the assumption that poor women organize solely for survival or that environmental analysis reveals women's subordination. Instead, the inclusion of women's exact words reframes grassroots women organizers as "dynamic, albeit under-valued, activists" (Jenkins 2008: 156). I draw from feminist conceptualizations of scale to stress the interconnections of globalized conservation and consumption processes that abruptly intersect in the lives of grassroots women and waste-management activists in coastal Yucatán. The women's ecological oral histories show how the mobilization of grassroots women's organizations is based on local, embodied experience, but also have direct regional influence for addressing garbage problems linked to globalization, increased consumerism, and global environmental change.

In focusing on women's narratives of change, I am not letting state or private organizations "off the hook" for their roles in normalizing specific conservation and development practices and policies that exclude women, namely by financing men's work in fishing and eco-tourism activities. At this point in time, while plastics consumption and waste are increasing from local residents and tourism, urban services continue to be neglected by many levels of government. Women have thus taken into their own hands the responsibility to maintain clean and healthy living spaces (Figure 9.3). The women's histories express their motivations for participating in urban sustainability work that was not being done otherwise in a region so highly focused on global concerns for tropical coastal conservation. The histories "further establish women's continued and important roles in producing creative spaces for the negotiation of meanings, practices, and politics

Figure 9.3 Women raking up seaweed and removing garbage from the beach
Photo credit: Anne-Marie Hanson

within urban environmental governance and vulnerable coastal areas" in the context of global environmental change (Hanson forthcoming). As one member of Las Costeras told me, we all should remember the analogism of José Narosky (1975: 14): "while there exists the person who throws broken glass on the beach, there is *also* the person who bends down to pick it up" (translation mine).

Notes

1 Throughout this chapter "garbage" is used to encompass various types of non-organic solid wastes, including local and externally produced solid wastes. Over 70 percent of solid wastes accumulating in coastal Yucatán are comprised of different plastics. Other common types of garbage include scrap metals, glass or metal containers, and discarded fishing gear.

2 Profepa (Procuraduría Federal de Protección al Ambiente) is the Federal Environmental Protection Agency in charge of enforcing federal environmental laws and regulations.

References

Beall, J. 1997. Thoughts on poverty from a South Asian rubbish dump – Gender, inequality and household waste. *Ids Bulletin-Institute of Development Studies*. 28(3): \73–90.

——. 2006. Dealing with dirt and the disorder of development: Managing rubbish in urban Pakistan. *Oxford Development Studies*. 34(1): 81–97. doi: 10.1080/13600810500496087.

Buenrostro Delgado, O. and I. Israde. 2003. La gestión de los residuos sólidos municipales en la cuenca del Lago de Cuitzeo, México. *Revista Internacional de Contaminacion Ambiental.* 19(4): 161–169.

Cahill, C. 2007. The personal is political: Developing new subjectivities through participatory action research. *Gender, Place & Culture.* 14(3): 267–292. doi: 10.1080/09663690701324904.

Capurro, L., J. Euán, and J. Herrera. 2002. Manejo sustentable del ecosistema costero de Yucatán. *Avance y Perspectiva.* 21: 195–204.

Carney, J. 1993. Converting the wetlands, engendering the environment – The intersection of gender with agrarian change in the Gambia. *Economic Geography.* 69(4): 329–348.

Castree, N., P. Chatterton, N. Heynen, W. Larner, and M.W. Wright. 2010. Introduction: The point is to change it. *Antipode.* 41: 1–9. doi: 10.1111/j.1467-8330.2009.00713.x.

Church, J.A., N.J. White, R. Coleman, K. Lambeck, and J.X. Mitrovica. 2004. Estimates of the regional distribution of sea level rise over the 1950–2000 period. *Journal of Climate.* 17: 2609–2625. doi: http://dx.doi.org/10.1175/1520-0442(2004)017%3C2609:EOTRDO%3E2.0.CO;2.

Cob Chay, J. 2011. Acuíferos contaminados – Celestún necesita $6 millones para nueva red de agua. *Diario de Yucatán*, March 16.

—— and W. Ucán Chan. 2010. Rezago en puertos yucatecos. *Diario de Yucatán*, April 15.

Córdoba Azcárate, M. 2006. Between local and global, discourses and practices: Rethinking ecotourism development in Celestún, Yucatán, México. *Journal of Ecotourism.* 5(1): 97–111.

Creel, L. 2003. Ripple effects: Population and coastal regions. Report. Washington, D.C.: Population Reference Bureau. 8pp.

Curiel, O. 2007. Crítica poscolonial desde las prácticas políticas del feminismo antirracista. *Nómadas.* 26: 92–101.

DOF (Diario Oficial de la Federacíon). 2004. Especificaciones de protección ambiental para la selección del sitio, diseño, connstrucción, operacíon, monitoreo, clausura y obras complementarias de un sitio de disposición final de residuos solidos urbanos y de manejo especial. In *NOM-083-SEMARNAT-2003.* México, D.F.: Semarnat.

——. 2011. Ley General del Equilibrio Ecológico y la Protección al Ambiente. Diario Oficial de la Federación. México, D.F.: Secretaría de Gobernación.

Elmhirst, R. 2011. Introducing new feminist political ecologies. *Geoforum.* 42(2): 129–132. doi: 10.1016/j.geoforum.2011.01.006.

Endres, D. 2011. Environmental oral history. *Environmental Communication.* 5(4): 485–498. doi: 10.1080/17524032.2011.610810.

Fanshawe, T. and M. Everand. 2002. The impacts of marine litter. Report of the Marine Litter Task Team. Marine pollution monitoring management group, Department of Environment, Bristol, UK. 43pp.

Fergutz, O., S. Dias, and D. Mitlin. 2011. Developing urban waste management in Brazil with waste picker organizations. *Environment and Urbanization.* 23(2): 597–608.

Fraga, J. 2004. Los habitantes de la zona costera de Yucatán: Entre la tradición y la modernidad. In *El Manejo Costero en México*, eds E. Rivera Arriaga, G. Villalobos Zapata, I. Adeath, and F. Rosado May, 497–506. Campeche: Universidad Autónoma de Campeche.

——. 2006. Local perspectives in conservation politics: The case of the Ria Lagartos Biosphere Reserve, Yucatán, Mexico. *Landscape and Urban Planning.* 74(3–4): 285–295.

——. 2009. Atrapados en un torbellino de cambios: La experiencia de las pesquerias tradicionales en las reservas marinas del Estado mexicano del Yucatán muestra la influencia de los factores socioeconomicos. *Revista Samdura*. 52: 26–31.

Fredericks, R. 2009. *Doing the dirty work: The cultural politics of garbage collection in Dakar, Senegal*, PhD dissertation, University of California, Berkeley, CA.

García de Fuentes, A., M. Xool Koh, J. Euán Ávila, A. Munguía Gil, and M.D. Cervera Montejano. 2011. La costa de Yucatán en la perspectiva del desarrollo turístico. In *Coleccíon Corredor Biológico Mesoamericano México: Conocimientos, Acciones, Diálogos*, ed. D. Becerra. México, D.F.: Comisión Nacional para el Conocimiento y Uso de la Biodiversidad.

Garcia-Frapolli, E., G. Ramos-Fernandez, E. Galicia, and A. Serrano. 2009. The complex reality of biodiversity conservation through Natural Protected Area policy: Three cases from the Yucatán Peninsula, Mexico. *Land Use Policy*. 26(3): 715–722.

Gold Bouchot, G., C.J. Montero, O. Zapata Pérez, J. Herrera Silveira, D. Pech Pool, L. Alpuche, and V. García Ríos. 2011. Evaluación de la condición ambiental de la Ría de Celestún, Campeche-Yucatán. Reporte Final. Mérida, Yucatán: CINVESTAV Unidad Mérida.

Gower, J. and S. King. 2011. Distribution of floating Sargassum in the Gulf of Mexico and the Atlantic Ocean mapped using MERIS. *International Journal of Remote Sensing*. 32(7): 1917–1929.

Gutberlet, J. 2008. *Recovering resources – recycling citizenship: Urban poverty reduction in Latin America*. Aldershot: Ashgate.

Haenn, N. 2005. *Fields of power, forests of discontent: Culture, conservation, and the state in Mexico*. Tucson, AZ: University of Arizona Press.

Hanson, A.-M. forthcoming. Women's oral histories of recycling and develpoment in Celestún, Mexico. *Gender, Place & Culture*.

Harris, L. 2006. Irrigation, gender, and social geographies of the changing waterscapes of southeastern Anatolia. *Environment and Planning D: Society and Space*. 24: 187–213.

Hazra, T. and S. Goel. 2009. Solid waste management in Kolkata, India: Practices and challenges. *Waste Management*. 29(1): 470–491.

Huysman, M. 1994. Waste picking as a survival strategy for women in Indian cities. *Environment and Urbanization*. 6(2): 155–174.

INEGI. 2010. *Censo de Población y Vivienda 2010*. México, D.F.: Instituto Nacional de Estadística y Geografía.

Jarosz, L. 2011. Nourishing women: Toward a feminist political ecology of community supported agriculture in the United States. *Gender, Place & Culture*. 18(3): 307–326. doi: 10.1080/0966369x.2011.565871.

Jenkins, K. 2008. Practically professionals? Grassroots women as local experts – A Peruvian case study. *Political Geography*. 27(2): 139–159. doi: 10.1016/j.polgeo.2007.07.006.

Kesby, M. 2005. Retheorising empowerment-through-participation as a performance in space: Beyond tyranny to transformation. *Signs: Journal of Feminist Theory*. 30(4): 2037–2065.

Leatherman, T.L. and A. Goodman. 2005. Coca-colonization of diets in the Yucatán. *Social Science & Medicine*. 61(4): 833–846. doi: 10.1016/j.socscimed.2004.08.047.

Lliteras, S.E. 2009. La basura, amenaza para la salud y el turismo en Celestún. *Revista de Yucatán*, July 21.

McGranahan, G., D. Balk, and B. Anderson. 2007. The rising tide: Assessing the risks of climate change and human settlements in low elevation coastal zones. *Environment and Urbanization*. 19(1): 17–37. doi: 10.1177/0956247807076960.

Mcllgorm, A., H.F. Campell, and M.J. Rule. 2011. The economic cost and control of marine debris damage in the Asia-Pacific region. *Ocean & Coastal Management*. 54: 643–651.

Medina, M. 2005. Serving the unserved: Informal refuse collection in Mexico. *Waste Management & Research*. 23(5): 390–397.

——. 2010. Solid wastes, poverty and the environment in developing country cities. Challenges and opportunites. Helinski, Finland: UNU World Institute for Development Economics Research (UNU-WIDER).

Mehra, R., T.T.N. Du, N.X. Nghia, N.N. Lam, T.T.K. Chuyen, B.A. Tuan, P.G. Tran, and N.T. Nhan. 1996. Women in waste collection and recycling in Ho Chi Minh City. *Population and Environment*. 18(2): 187–199.

Mohanty, C. 2003. *Feminism without borders: Decolonizing theory, practicing solidarity*. Durham, NC: Duke University Press.

Moore, S.A. 2008. The politics of garbage in Oaxaca, Mexico. *Society & Natural Resources*. 21(7): 597–610.

——. 2009. The excess of modernity: Garbage politics in Oaxaca, Mexico. *Professional Geographer*. 61(4): 426–437.

Morton, R.A., T.L. Miller, and L.J. Moore. 2004. National assessment of shoreline change: Part 1 historical shoreline changes and associated coastal land loss along the U.S. Gulf of Mexico. St. Petersburg, FL: U.S. Department of the Interior U.S. Geological Survey.

Myers, G. 2005. *Disposable cities: Garbage, governance and sustainable development in urban Africa*. Burlington, VA: Ashgate.

Narosky, J. 1975. *Si todos los hombres*. Buenos Aires: Ediciones Antartida.

NAU. 2005. *Ecological oral histories collection*. Flagstaff, AZ: Northern Arizona University, Cline Library.

Nightingale, A.J. 2011. Bounding difference: Intersectionality and the material production of gender, caste, class and environment in Nepal. *Geoforum*. 42(2): 153–162. doi: 10.1016/j.geoforum.2010.03.004.

NOAA. 2014. *Marine debris program*. Available from http://marinedebris.noaa.gov/. Accessed March 16, 2013.

Notimex. 2013. Reportan que 70% de cenotes en Yucatán están infestados. *Notimex*, June 16.

Pacheco, Y. 2010. Limpian de PET la costa de la Península de Yucatán. *SIPSE noticias*, November 19.

Pratt, G. and S. Hanson. 1994. Geography and the construction of difference. *Gender, Place & Culture*. 1(1): 5–29. doi: 10.1080/09663699408721198.

Radel, C. 2012. Gendered livelihoods and the politics of socioenvironmental identity: Women's participation in conservation projects in Calakmul, Mexico. *Gender, Place & Culture*. 19(1): 61–82.

Robertson, H.A. and T.K. McGee. 2003. Applying local knowledge: The contribution of oral history to wetland rehabilitation at Kanyapella Basin, Australia. *Journal of Environmental Management*. 69(3): 275–287. doi: 10.1016/s0301-4797(03)00155-5.

Rosado, R.M. 2009. Pepenadoras y educación ambiental: por una cartografía de la basura. *Utopía y Praxis Latinoamericana*. 14(44): 109–115.

Scanlan, J. 2005. *On garbage*. London: Reaktion Books.

SEDESOL and Oportunidades. 2013. *Funciones de oportunidades*. Available from www.oportunidades.gob.mx/Portal/wb/Web/funciones_de_oportunidades. Accessed March 16, 2013.

SEDUMA. 2009. Programa especial para la prevención y gestión integral de los residuos 2009–2012. Mérida, Yucatán: Secretaría de Desarollo Urbano y Medio Ambiente.

SEMARNAT. 2008. Programa regional de educacion para la sustentabilidad en âreas naturales protegidas (2008–2012). Región Peninsula de Yucatán. Mérida, Yucatán: Secretaría del Medio Ambiente y Recursos Naturales.

——. 2009. Programa nacional para la prevencion y gestion de residuos. México, D.F.: Secretaría del Medio Ambiente y Recursos Naturales.

Sepulveda, P. 2013. Celestún y Río Lagartos: focos rojos en destrucción manglar. *El Milenio*, May 11.

Smith, N. 1984. *Uneven development: Nature, capital and the production of space*. Cambridge, MA: Basil Blackwell.

Sultana, F. 2009. Fluid lives: Subjectivities, gender and water in rural Bangladesh. *Gender, Place & Culture*. 16(4): 427–444. doi: 10.1080/09663690903003942.

Truelove, Y. 2011. (Re-)conceptualizing water inequality in Delhi, India through a feminist political ecology framework. *Geoforum*. 42(2): 143–152. doi: 10.1016/j.geoforum.2011.01.004.

UN-HABITAT. 2008. Annual report. United Nations Human Settlements Programme. Nairobi: UN-HABITAT. http://unhabitat.org/. Accessed July 6, 2014.

UN-OCEANS. 2013. Human settlements on the coast. In *UN atlas of the oceans*. N.p., n.d. www.unoceans.org/. Accessed June 12, 2014.

UNEP. 2005. Marine litter: An analytical overview. Nairobi: United Nations Environment Programme.

——. 2006. *Marine litter in the wider Caribbean*. Kingston: UNEP Caribbean Environment Program.

UNEP-CAR/RCU. 2008. Marine litter in the wider Caribbean region: A regional overview. United Nations Environment Programme.

Vásquez Avila, L.A., S. Ota, and C.M. Alarcón Lazcano. 2011. *Manual para el establecimiento de un sistema de manejo de residuos sólidos en ciudades, pequeñas, comunidades rurales y areas naturales protegidas. Estudio de caso: Municipio de Celestún, Yucatán, México proyecto de conservación de humedales en la peninsula de Yucatán*. Mérida, México: CONANP-JICA.

Vazquez-Dominguez, E. and H.T. Arita. 2010. The Yucatan peninsula: Biogeographical history 65 million years in the making. *Ecography*. 33(2): 212–219.

Wiles, J.L., M.W. Rosenberg, and R.A. Kearns. 2005. Narrative analysis as a strategy for understanding interview talk in geographic research. *Area*. 37(1): 89–99.

10 Storytelling water north of the future *Héen Kas'él'ti X̱oo*[1] (among the ragged lakes)

Collaborative water research with Carcross/Tagish First Nation, Yukon Territory, Canada

Eleanor Hayman in collaboration with Mark Wedge/Aan Gooshú *and Colleen James*/G̱ooch Tláa

Introduction

> When our ancient people talked about water, what the western world calls H_2O, they would say 'Haa daséigu a tóo yéi yatee': Our Life is in the water . . . Our breath is in the water.
>
> (Elder David Katzeek/*Kingeistí*, Shangukeidí Clan of Klukwan, Eagle Moiety, November 2013, personal communication)

This chapter is about water. Not bottled water or the chemical formula H_2O, but water understood and celebrated within the Tlingit language and Tlingit/Tagish culture. By extension it is also a commentary on the dominant vision of water that is gendered, narrow, and now naturalized within global water governance rhetoric—a 'water' often referred to as 'modern water.' A Tlingit relationship with water can be traced through over 10,000 years of Tlingit oral tradition (in particular narrative storytelling) but is also reflected in Tlingit place names around the Southern Yukon Lakes, where over three-quarters of place names are water-related. In a Tlingit cosmology, stories emerge from and are co-dependent with ecological processes, something that I call 'slow activism.' How might 'slow (and therefore radical) activism'—a powerful and highly sustainable form of environmental storytelling—act as both a counterpoint to dominant water rhetoric as well as a means of resistance to essentialized readings of water? Indeed, who gives voice to water in the context of water management? How do the framings, possession and control of water, and its circulations in and out of social spaces undermine or promote particular worldviews?[2] These questions are gaining new and urgent momentum in this—so-labeled—epoch of the Anthropocene and its associated politics of climate change. In the circumpolar north where global warming is providing a catalyst for nations hungry for 'untapped resources,' these questions are of unprecedented urgency (Fox 2014).

The ecological crisis that Earth currently faces forces a critical analysis of some of the most fundamental assumptions within the dominant culture—that is, the conceptual and imagined space of western, patriarchal, capitalist society. Many indigenous intellectuals, feminist philosophers, environmental historians, cultural anthropologists, and science technologists are of a similar voice—that there is a requirement for a 'revolution of the mind,' a 'new consciousness,' a 'paradigm shift,' and a 'reconfiguration of western thought' to deal with and counteract this ecological crisis and the specter of global climate change (Code 2006; M. Wedge, 2014, personal communication). One of the most convincing tools to affect this shift is also a fundamental assumption of this chapter: that there is not one single reality defined by the assumptions of the dominant culture, but rather multiple realities that can be known. Māori indigenous scholar Linda Tuhiwai Smith elaborates with these words:

> The globalization of knowledge and Western culture constantly reaffirms the West's view of itself as the center of legitimate knowledge, the arbiter of what counts as knowledge and the source of 'civilized' knowledge. This form of global knowledge is generally referred to as 'universal' knowledge.
>
> (Smith 1999, 63)

Simply put, the assumption of this collaborative research is that there is not a uni-verse but a pluri-verse (Blaser 2010; Law 2010; Mignolo 2013). As one counter-story to the notion of the uni-verse and its assumptions about water and approaches to water 'governance,' and together with Carcross/Tagish First Nation (CTFN) community consultants Mark Wedge and Colleen James, I explore how Tlingit cultural practices and performances in relation to *héen* (water) are one of many critical pluri-verse realities (see Figure 10.1).

This chapter is primarily concerned with the ideological or structuring principles in cultures which present specific "constructions of gender, nature and the relationship between the two" (Agrawal in Bavington et al. 2004, 163). The first section introduces a Tlingit/Tagish cosmology and organizational structure, which is in a process of revitalization after over a century of colonial suppression. The second section identifies a problematic relationship between indigenous philosophy and notions of feminism(s) implicit within feminist political ecology, which itself is a product of particular 'western' historical circumstances. In the third section, I synthesize the idea of 'modern water' and argue that it is stitched into petrified dualisms (i.e. nature/culture, wild/civilized, male/female) that exist both as relics of a particular colonial history and now as a normalized way of understanding water. The final sections offer a Tlingit worldview as a counterpoint to dominant and essentialized understandings of water. This chapter writes a Tlingit voice into a common environmental water (hydro-commons) history and fundamentally enriches and adds alternatives in moral debates on water ethics and approaches to water 'governance.'

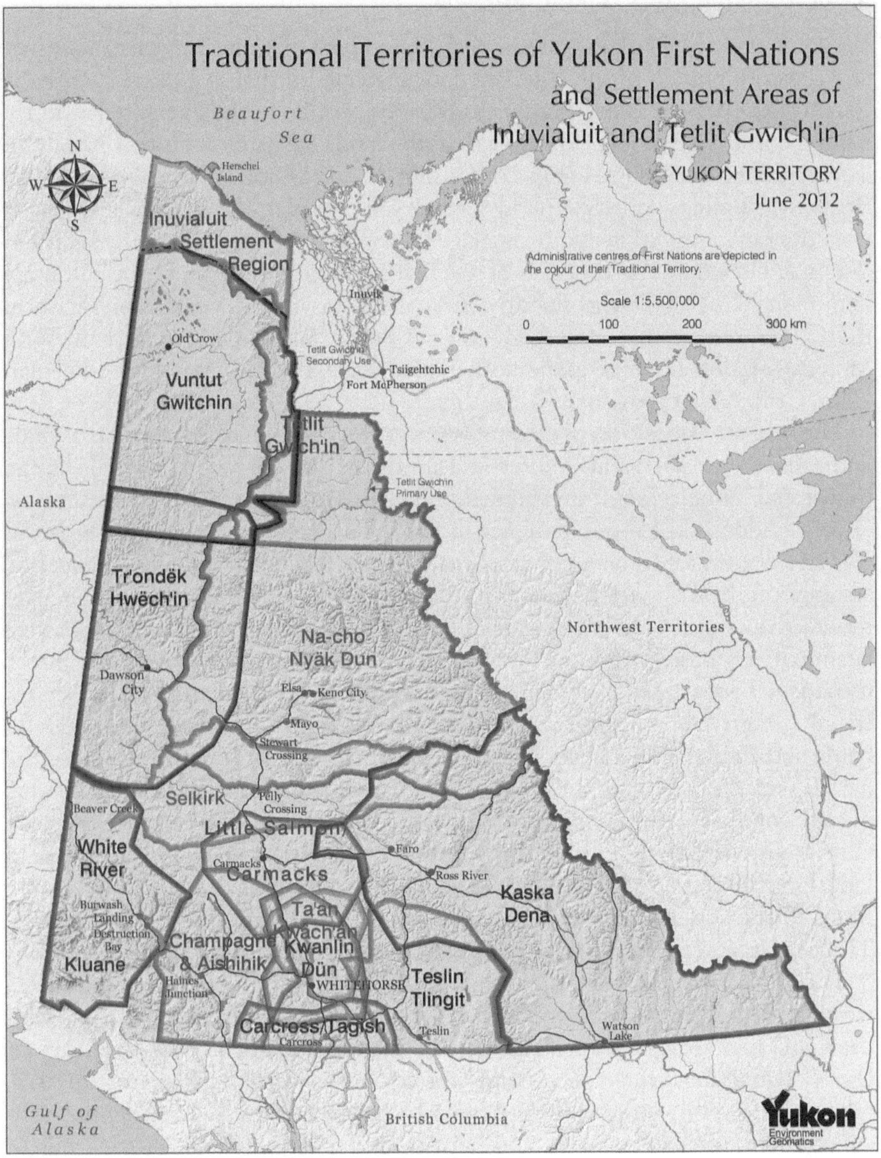

Figure 10.1 Location of research: Carcross/Tagish First Nation traditional territory, Yukon Territory, Canada

Source: Environment Yukon. To view the original color map, go to www.environmentyukon.ca/maps/view/detail/1/10/74/ (accessed October 10, 2013)

Tlingit/Tagish culture and language

Tlingit is a native Alaskan and southern Yukon First Nations' language, a distinct and separate branch of the Na-Dené language family that is spoken fluently by roughly 200 Alaskan Native and First Nations peoples in the circumpolar north. Tlingit culture is far more extensive than the language itself, defining 130 Tlingit Tribes, Clans, and Clan Houses comprising 17,000 people[3] across Alaska (US), British Columbia, and the Yukon Territory (Canada). The Tagish language is from the northern Athapaskan language family no longer spoken in the southern Yukon or elsewhere. Descendants of the Tagish[4] language also spoke the culturally dominant Tlingit language fluently. Two of the last fluent speakers of Tagish in the CTFN community were Lucy Wren/*G̲ooch Tláa* (Dak̲l'aweidí Clan, Wolf Moiety)—Colleen James' grandmother, and Angela Sidney/*Stóow* (Deisheetaan Clan, Crow Moiety)—Mark Wedge's aunt.

The violent imposition of colonialism with its patriarchal values such as male authority over women and children and control of property wrought havoc on the traditional Tlingit/Tagish matrilineal and egalitarian clan system. This is most clearly evidenced by the Indian Act of 1876. This piece of Canadian legislation was and continues to be a powerful tool for the federal government, reflecting assumptions held about indigenous peoples by the dominant Euro-Canadian, late nineteenth-century society (Coates 2008, 1–2). The Indian Act's core aims were 'assimilation, protection, and civilization' of 'aboriginal' communities into a Canadian system. The Act defined who and how an individual was a 'status' Indian. Often referred to as the 'double-burden,' the Act marginalized women for being both Indian and female. By redefining status, the Act further destroyed the matrilineal and egalitarian clan structures with the assumption that patriarchal nuclear families were the 'natural' model for all cultures. The Act rejected clan matriarchs, outlawed cultural and spiritual ceremonies, and placed control of lands and resources with the Department of Indian Affairs. However, despite repeated and systematic suppression and assimilation by European colonizers, the Tlingit clan and kinship system continues to show astonishing resilience.

The long road to self-government for many Yukon First Nations began in 1973 with Elijah Smith and a delegation of Yukon Chiefs starting Land Claims with the Canadian Government based on the historic document *Together Today for Our Children Tomorrow*. The *Umbrella Final Agreement*, finally signed in 1993, reflected 20 years of gruelling negotiations and addressed aboriginal self-government, settlement lands, and hunting and fishing rights. Self-government for many Yukon First Nations has gone a long way to reverse the colonial process, especially as the Indian Act 'ceases to apply' with the signing of a Final Agreement with the Canadian Government. Many indigenous traditional practices, structures, values, ceremonies, and authorities have been revived. However, land is no longer clan owned but under the collective control of each First Nation government.

The Tlingit clan system is rooted within two moieties or halves—Wolf and Crow—and provides the fundamental balancing force and point of reference in

Tlingit society, in particular CTFN (M. Wedge, 2013, personal communication). Gender is not significant in terms of hierarchies or privileges within Tlingit society. Rather, individual recognition and a deep consideration of each person's virtues, attributes, and achievements are highly valued and reflected in Tlingit language, philosophy, and history. Tlingit/Tagish women continue to hold as many strong leadership positions as men. Historically, Tlingit women were primary educators for language and culture, primary negotiators in trade agreements, and often in charge of trade finances. Gender does not delineate status or power. One example demonstrating that gender has a fundamentally different role in Tlingit society than in colonial society is shown within the Tlingit language. Simply put, the third-person pronoun that refers specifically to humans is ungendered. Linguistically, this introduces another complex historical/conceptual construction of how gender can be imagined. Together with a rich and powerful oral tradition of storytelling, these linguistic, cultural, social, and spiritual elements form the strong philosophical basis of Tlingit identity and sense of place.

It is readily recognized by many individuals in First Nation communities that they are as much a part of the patriarchal capitalist system of mastery and exploitation as they are an intimate part and advocate of a matrilineal Tlingit clan culture. This is critical to acknowledge as it problematizes the tendency to classify knowledge into either indigenous or western (Agrawal 1995). First Nations themselves work within dualistic rhetorical frameworks such as Traditional Ecological Knowledge (TEK) and 'western' scientific knowledge. Yukon First Nations and CTFN in particular cannot be essentialized into either 'indigenous' or 'western' categories, categories that are themselves a product of a particular way of viewing the world.

This chapter draws extensively on collaborative ethnographic water research undertaken between 2012 and 2014 with the inland Tlingit/Tagish CTFN community whose traditional territory embraces the Southern Yukon Lakes in the Yukon, Canada; the headwaters of the Yukon River. The villages of Carcross and Tagish were well established before first contact with European traders in the 1800s. CTFN is one of 14 First Nations in the Yukon Territory, 11 of which (including CTFN) are now self-governing. CTFN signed their Final Agreement, a legal treaty that designates settlement lands and deals with economic, wildlife, natural resources, and heritage issues, with the Federal and the Yukon Government on October 22, 2005.[5] The first Act drawn up by the CTFN Government was a Family Act rooted in Tlingit oral traditions and its defining value system. This collaborative water research is a potential framework to develop legislation for a CTFN Government Water Act, located as the Family Act is, in the fundamental teachings of over 10,000 years of Tlingit/Tagish narrative storytelling. Tlingit oral narratives reveal what could be considered as one of the first sets of environmental (water/*héen*) ethics and philosophy that is inherently part and parcel of Tlingit values embodied in Tlingit oral tradition.

Indigenous women and feminism: decolonizing feminist political ecology?

There is growing scholarship and sophisticated critique by indigenous scholars on the nature, epistemology, and colonial history of 'feminism.' Implicit within this scholarship is a theorizing of indigenous feminism(s)/native feminism(s) that includes a healthy amount of skepticism and rejection of 'western' feminism historically located at the nexus of patriarchy, colonialism, and white supremacy (Smith 2011). First and second wave 'western' feminist movements were generally catalyzed by white, middle-class academics, and native women's concerns were not addressed. Furthermore, European-based cultural labeling, constructions of gender, and notions of feminism are simply that—constructions and terms reflecting European frameworks and generally all in the dominant English language. Therefore feminism is widely regarded as an intimate part of the colonial project by many Canadian First Nations/aboriginal peoples (B. Sembsmoen, 2013, personal communication; Grey 2004, 12). With this background it is no surprise that the academic field of feminist political ecology is troubled by the notion that ideas of feminism, indigeneity, women, and gender do not sit comfortably together.

My collaborative water research with CTFN centers on the Tlingit word *héen* (water). *Héen* is a powerful destabilizing and de-familiarizing setting for feminist political ecology. In contrast to the blanket term 'feminism,' feminist political ecology is far more nuanced with its central concerns about power and relationships. While definitions of feminist political ecology certainly draw on many feminist approaches and perspectives, feminist political ecology itself claims a particular sensitivity to situated knowledges, or knowledges born out of specific spatial and embodied practices (Elmhirst 2011, 130). In other words, feminist political ecology has a particular set of competencies for tracing, tracking, interrogating, and unsettling dominant assumptions about relationships between gender and water and resulting water governance ideologies and structures. In particular, it places indigenous knowledge outside of a rather narrow and restrictive anthropological context to "encourage a respectful examination of local knowledges within the global context" (Turk 2011, 57). While it is clear that a decolonization of knowledge—specifically gendered and indigenous knowledges within feminist political ecology—must be further problematized theoretically, collaborative research does offer rich opportunities for the very nature and intent of feminist political ecology to be "interrupted and challenged" (Cruikshank 1998, 165). Indigenous scholars, for example, are vocal in concerns about the (post)colonial re-ordering of conceptual space, the essentializing tendencies of western approaches to indigenous knowledges/cultures, and the imposition of sets of gendered roles on indigenous communities (enforced in a Canadian context by the Indian Act of 1876). These categorizations can be read as sophisticated forms of continued cultural colonization.

The tension between these concepts of indigenous women and feminism is voiced by Beverly (Bev) Sembsmoen/*Shooaateen* (Dak̲l'aweidí Clan, Wolf Moiety,

CTFN) at Millhaven Bay on the shores of Lake Bennett, Southern Yukon, in August 2013. Bev has worked for the CTFN Government since 1996 and was a key technician for CTFN's Final Agreement with the Yukon Government. She considers herself an activist and not a feminist, something she elaborated on in an interview when she stated that she takes issue with the word feminism, arguing that it is conceptually too narrow. Bev illustrates this with the recent protest movement *Idle No More* and alludes to both the movement's non-hierarchical structure, its female driven momentum, its focus on responsibility for the more-than-human world, and its location in decolonizing discourses:

> That's what's going on with Canada right now with Idle No More. It is the women of Canada who are standing up and telling the government "We've had enough!". . . . Well now the aboriginal people in Canada are rising up and saying "Enough! We're not second class citizens in our own country. If you guys don't like us, you go home because we're already home." Is that feminism? No it's not. It's women without a title. Where it's just mothers and grandmothers and great-grandmothers and daughters. We're all saying it's our job to protect mother earth. It's our job to protect families, and water is a critical life source . . . women are protecting their water sources.

Idle No More began as a reaction to an impending Canadian parliamentary omnibus bill C-45 (now a law) whose perceived threat was that it would erode both indigenous sovereignty and environmental protection laws on navigable waterways. The intersection of water protection with predominantly female activists offers a particular vision of indigenous women's relationship with water, but not one that can necessarily be pigeonholed into a particular way of knowing. Activism in the form of ecological and/or linguistic protection and regeneration is most often a function of family circumstances in First Nations communities in Canada that reflect a combination of impacts from the 1876 Indian Act, linguistic suppression in residential schools, and the maintenance (or not) of cultural practices that include hunting, trapping, and fishing. Colleen James summarizes:

> There are only basic clan laws and oral histories passed down to my generation, a gift not all of us had benefit of. No old people in family, sickness took many, mission school, religion, government, new laws, all played a role in the demise, breakdown of traditional knowledge. Today we search the ashes of the fire, hoping for remnants, artefacts that will help us remember who we really are.[6]

Many Elders of Tlingit/Tagish descent are knowledge keepers—gifted storytellers who practice Tlingit/Tagish narrative traditions. A high proportion of storytellers are female, but stories themselves are not gender-specific. Men and women storytellers do, however, often stress different aspects within stories. For example, female Elders have tended to use narrative storytelling to demonstrate a storied connection to the land and waters, to illustrate how particular waters are mnemonics for particular clan histories, claims, and narratives, and how narrative storytelling itself mimics the circulatory, replenishing, sustaining nature of water. Male storytellers on the other hand tend to focus more on the imposition of 'modern water' on traditional water-related material culture and identity, offering highly detailed descriptions of the history, fluctuations, and impacts of the changing hydrology and ecology of the Southern Yukon Lakes region.

It becomes clear then that the problematic intersection of feminism(s), females/males, and indigeneity is further complicated by the themes of cultural identity, nationalism, and decolonization. This is most recently and vigorously discussed by indigenous scholars in *Indigenous women and feminism* (Suzack et al. 2010) who raise the question, "Can feminism inhabit discourses that marginalize the question of gender?" I argue that the most recent literatures on feminism(s), the politics of indigenous feminism(s), and political ecology during most of the early twenty-first century shows that it can. This 'feminist' literature "extends well beyond female-feminist concerns to engage with wide-ranging social-justice and politics-of-knowledge issues of dominance and oppression" (Code 2006, 14). More recent critical enquiry on the coloniality of knowledge (Leff 2012) echoes Mark Wedge's concern that dominant systems of governance and (water) management in Canada do not reflect indigenous ways of knowing. Enrique Leff writes that "decolonization of knowledge leads to inquire how Eurocentric ideas – from Greek philosophy to modern science and technology – were introduced to traditional societies and cultures through conquest, colonization and globalization, invading indigenous modes of thinking and their cultural lifeworlds" (2012, 7). Within feminist political ecology circles, Mollett and Faria have contributed to this debate by arguing that "the decentering of gender remains unfulfilled in feminist political ecology and that a more complex and messier, notion of 'gender' must be recognized" (2013, 116). This insight opens up spaces for dialogue, and clearly recognizes that feminist political ecology requires more rigorous problematization within the context of decolonizing gendered and indigenous knowledges.

The following section illustrates how 'modern water' is a complex blend of gendered and historical narratives, persuasive in defining and influencing global level discourses about water control, management, and governance. This notion of a passive, yielding (feminine) water has been constructed with the ideological footprint that it *needs* to be managed (Bavington et al. 2004). The critical interlacing of colonial ideas about water, gender, purity, and power has suppressed other water worldviews (Hayman 2012). A Tlingit vision of water is rooted in 'baselines' that embrace relationships, reciprocity, and respect. If there is any 'management' talk, it is about managing communities in relation to the dynamics and rhythms of water's flows, floods, and circulations.

Modern water

'Modern water'[7] is a complex blend of gendered and historical narratives that have been persuasive, violently or otherwise, in defining and influencing at the global level, discourses about water control, management, and governance. Water became increasingly invisible and abstracted with the technological manipulation of water and urban water infrastructure provision in western European cities in the nineteenth century (Linton 2010, 2014). Greta Gaard (2001) makes the further link between the positions and treatment of women in western culture and the treatment of nature (water).

In 'Environmental orientalisms' (2000) Sawyer and Agrawal trace how this thinking further exposes a form of labeling within the colonial imagination which ossifies the gender/water/race nexus when they write: "native topographies and peoples [were labeled] as feminine spaces to be violated, and thereby instantiated a sexual/racial hierarchy between colonizer and colonized" (2000, 72). Donald Worster's concept of 'imperial water' illuminates the hydrological violence imposed on pre-modern waters in the form of the western hydrological discourse (Worster in Linton 2010). 'Imperial water' is intimately linked to Sawyer and Agrawal's narrative of gendered and sexualized 'virgin waters,' and highlights how, through a reading of both water *and* gender, fractures across new lines of race, class, and ethnicity are illuminated. One could describe this as *hydro*-orientalism: "by bringing colonial waters under a western epistemological and material control" (Linton 2010, 59), colonial powers forced the development and diffusion of the ontology of modern water on lands and waters they 'discovered.'

Challenging the dominant water story: storytelling water north of the future

> If we start drinking that water [bottled water] I think we're starting to forget who we are.
>
> (Elder Norman James/*K̲aaklig̲é*, Dak̲l'aweidí Clan, Wolf Moiety, CTFN, August 2013)

> The essence of being Tlingit is what comes in exploring – that name – *héen*.
>
> (Elder David Katzeek/*Kingeistí*, Shaŋgukeidí Clan of Klukwan, Eagle Moiety, November 2013, personal communication)

In this section, I explore three ways relationships with water are framed within Tlingit culture. They reveal and challenge cultural identity, nationhood, and spirituality, and are framed within the circulation of water in and out of human bodies and the body of social space itself.

Elder Norman James/*K̲aaklig̲é* (Dak̲l'aweidí Clan, Wolf Moiety) grew up in and around Carcross and Tagish. He is one of four fluent Tlingit speakers in the CTFN community. His words at the beginning of this section highlight the historical consciousness of the colonial project[8] and the current links and concerns with distant cultures of consumption. Colleen James illustrates how relationships with water and ice are shifting as a result of warmer temperatures, the endangered

Tlingit language, and Tlingit cultural revitalization projects. She takes the practice of packing (carrying) water for daily use as a critical component in closely observing ecological patterns and rhythms. Elder Mark Wedge concludes with a commentary on the critical importance of narrative storytelling, what stories do, how they can behave and their evolving social life and function—all essential aspects in understanding 'storytelling water.' The selected narratives act as a counterpoint to dominant utilitarian and managerial attitudes towards water.

I situate these narratives within the context of four 'modern water' dialogues currently taking center stage in the Yukon Territory. First, the Yukon Government has recently released (June 11, 2014) a Yukon Water Strategy that involved a two-year process of community consultations.[9] Second, the CTFN Executive Council and General Council banned potential hydraulic fracking projects in November 2013, reinforced by a submission for a 'Frack-Free Traditional Territory' to the Yukon Legislative Assembly Select Committee Regarding the Risks & Benefits of Hydraulic Fracturing in September 2014.[10] As CTFN *K̲aa Shaadé Héni* (Chief) Danny Cresswell (Deisheetaan Clan, Crow Moiety) explains, "It's our responsibility to protect the environment and our Traditional Territory, which includes the headwaters of the Yukon River."[11] Third, the Yukon River Inter-Tribal Watershed Council's Yukon River Watershed Plan (approved August 2013) is built around a set of measurable Water Quality Standards, but is also explicit about the role of First Nation and Tribal governments in Yukon River governance. Lastly, the most significant aspect of modern water's ideology for CTFN is the current proposal by local energy provider Yukon Energy to regulate the Lewes Dam at the exit mouth of Marsh Lake. The proposal would keep the Southern Lakes (Bennett, Nares, Tagish, and Marsh Lakes) level artificially raised each autumn to provide extra electricity throughout the winter for the Yukon Territory via the hydroelectric dam at Whitehorse.

CTFN's historical relationship with the Southern Lakes and headwaters of the Yukon River is grounded in an indigenous notion of respect. This relationship might well be regarded historically in this region as one of the first water ethics. Colleen James describes it like this:

> When you're traversing this land on the water and on the river it's really important to have respect for it like the old people say. That water gives us life and gives everything on this planet life. Also, it can take it just like that in the blink of an eye, in the turn of the wind.

Water, respect, and Tlingit identity are closely related. Just as water and blood circulate with similar verb forms in the Tlingit language, so too must respect. Langdon (2013)[12] points out that this circulation is a fundamental prerequisite for existence: "It is through the continuous circulation of respect – in thought and deed – exhibited in connections and fulfillment of obligations in various socially and ritually prescribed ways that the Tlingit pursue a morality that will insure the continuity of existence." This statement reiterates Elder David Katzeek's words at the Tlingit Clan Conference in Juneau, Alaska in November

2013, "Our life is in the water; our breath is in the water." Water becomes literally and metaphorically the master verb within a Tlingit cosmology that is itself a highly complex and sophisticated web of spiritual connections and obligations.

The Southern Lakes were and still are to an extent the fundamental matrix for First Nation's everyday life. From transportation routes, to fishing for various types of fish (such as Whitefish, Arctic Grayling, Lake Trout, Herring, and Northern Pike), to drinking water, the Southern Lakes are seamlessly bound up in First Nations' cultural practices. Although not coastal, the Southern Lakes have their own particular 'tidal' rhythm (M. Wedge, 2014, personal communication) to which cultural practices mimic and respond. For example, the glacial and snow-fed Southern Lakes reach their peak water level in July/August responding to a combination of glacier and snowmelt at high latitudes (Benkert and Johnson 2011, 4). Traditionally this was the start of the Tlingit year[13] when the first of the salmon runs took place. The freeze and break-up (of ice) on the Southern Lakes also provide ice-bridges that are fundamental in making trapping and other culturally important hunting activities possible.

However, social relationships and understandings of ice in particular are changing as a result of shifting hydrological and climate patterns and an increasingly endangered Tlingit language (Figure 10.2). Colleen James confirms, "Ice (*t'éex'*) is behaving differently to what my ancestors told us to expect." Intimate knowledge of ice used to be a prerequisite for dog teams and ice fishing. As Colleen remembers, "My mother told me of Grandpa and how he would safely

Figure 10.2 Windy Arm on Tagish Lake, February 2013 ('Tagish' means the sounds of the break-up of ice in the Tagish language)

Photo credit: Heather Jones hpj Photography, www.hpjphotography.com

guide the whole family by dog team from Tushi River to Carcross every spring and fall." This knowledge is slowly being lost, as not "many of us have experienced ice life, something so important."

Tlingit language revitalization projects are critical as Tlingit is embedded with extraordinary amounts of precise, detailed ecological knowledge. Environmental change and cultural change in this case are complicated by the violent history of colonialism that remains systemic, evidenced perhaps most strikingly by the recent choreography and rhetoric of the so-called Anthropocene (Hayman forthcoming). Colleen James underlines the correlation between language, environmental knowledge, and change by explaining that having a limited knowledge of the Tlingit language: "[a]ffects relationships with water hugely, in that we are unable to describe water, water behavior, and various forms that water takes." This includes "many descriptions of ice, ice formations, ice sounds, pressure cracks, and air holes."

The very nature of traditional ecological knowledge, its relevance, its value, and its ability to adapt is being re-thought by many members of the CTFN community. This is a direct response to the agency of water; there is less snow (*dleit*) in the early winter months, thinner ice, and relatively more glacial runoff at peak flow periods in the Southern Lakes in August (M. Wedge, 2014, personal communication). Remote sensing satellite data on the Yukon River Basin appears to coincide with these detailed and close observational accounts, but lacks social sensitivities (Semmens et al. 2013).

Bev offers a more nuanced analysis of these impacts and how traditional ecological knowledge is understood: "Traditional knowledge is changing now with global warming. Knowledge about our water, about our animals, about the seasons . . . a lot of it is different now" (B. Sembsmoen, 2013, personal communication). Resilience is implicit within accounts of accelerated change driven by rising water temperatures, accelerating glacial melt, and unpredictable hydrological rhythms. Resilience in this case comes in the form of Tlingit '*Haa K̲usteeyí*' which is defined as 'accounts' or 'stories' in English. However, *Haa K̲ usteeyí* is more than that. It means 'all of it' and embraces *Haa kusun*, or 'the center of things.' As Colleen James describes it, "*Haa K̲usteeyí* is language! culture! food! water! family! community! land! animals! plants! universe! how to treat people! How we live and do things, all of what we know!"[14] It is where virtues, ethics, storytelling, and water meet and is part of the essence of what it means to be Tlingit.

Tlingit and storytelling

Narrative ecologies

> [Tlingit] storytelling is not about anchoring ideas or concepts. It is more about creative thinking, about circular thinking and is rarely coupled with any form of what academia calls 'analysis.'
>
> (M. Wedge, 2014, personal communication)

Storytelling is a profound and intrinsic part of life within Tlingit culture, even though many Elders lament its declining importance and performance among the younger generations. Storytelling as method in collaborative research has come under limited critique (Howitt and Suchet-Pearson 2006; Kovach 2009). Both Neil McLeod (2009) and Renate Eigenbrod (2014) propose that working with stories or thinking poetically "involves the movement away from the epistemological straightjacket and the colonial box that social sciences have often placed on indigenous narratives" (McLeod 2009, 109). Effectively, the positioning of stories highlights situated knowledges and connects disciplines by positioning participants "in the cultural, linguistic, and physical environments where each story is set" (Eigenbrod 2014, abstract). Storytelling resonates with the concept of the pluri-verse that counters the colonial project by actively and openly engaging in "multiple narratives with multiple purposes" (Howitt and Suchet-Pearson 2006, 59–60). Cruikshank adds to this conversation with the reflection that "an enduring value of informal storytelling is its power to subvert official orthodoxies and to challenge conventional ways of thinking" (1998, xiii). This is one of many functions of narrative storytelling. The power of strong stories is a form of both resilience and resistance to other colonizing narratives that flaunt certain ideologies. However, as Robert Bringhurst makes clear:

> Oral culture also means much more than telling stories. It means learning how to hear them, how to nourish them, and how to let them live. It means learning to let stories swim down into yourself, grow large in there and rise back up again.
>
> (2006, 175)

"How stories are performed, circulate, and play out their social lives is of fundamental importance in Tlingit culture" (M. Wedge, 2014, personal communication).

The imposition of and resistance to modern water's ideology is part of what I call 'narrative ecologies.' Narrative ecologies describe the dynamic social life of stories that reflect and co-evolve with ecological conditions. Stories themselves are sets of nested ecologies that have agency in unexpected and profoundly spiritual ways within Tlingit oral tradition. Stories chart and reflect both a particular idea of dwelling and reciprocity with other species, but also trace a profound relationship with a spirited animate world that is both spoken of and spoken to. The ongoing Tlingit dialogue with its various fluid ecologies and cartographies tracks an ancient understanding of dwelling that continues to have relevance today.

The following examples of the intersections of water with narrative storytelling are all offered or performed by female First Nations Elders. One aspect of their narrative ecologies can be witnessed in the archival record as early as 1915 before any technological structures (dams) were erected on the Yukon River. For example, at the 1994 International Storytelling Festival in Whitehorse, Jessie Scarff (an Elder from Kwanlin Dun First Nation) told the story of forced Indian settlement removal from waterfront sites along the Yukon River in Whitehorse.

Jessie used materials from the Yukon Archives in an ironic move (archives are where white people go to learn about history (J. Cruikshank, 2014, personal communication) to illustrate to her audience a deep rupture between the vision of two very different kinds of water—modern water and pre-modern (Tlingit) water. Jessie selected an extract from the local newspaper, the *Whitehorse Star* (1915): "It is better for the Indians that they should be away by themselves and it is certainly better for the town that they be not camped so close to the source of public water supply" (*Whitehorse Star*, October 22, 1915 in Cruikshank 1998, 152). In the article, the Yukon River is framed in terms of utilitarian, economic parameters. It is presented to the readers of the *Whitehorse Star* as primetime waterfront space that 'Indians' by their very presence devalue, and that Indians must also be located away from the source of public water supply. Racist dualisms of pure/impure reflect the fantasies and anxieties of a particular colonial imagination. This in turn constructed cultural identities and labeled specific waters in particular and troubling ways for the Tlingit and many other indigenous communities. The colonial labeling of water as 'urban' or 'waste,' for example, includes specific ideas and assumptions about its circulation and preferred control. This labeling is in complete contrast to assumptions embedded within water-related Tlingit place names. Many Tlingit dwelling places were chosen based on *ishkahéeni*—the availability of cold, oxygen-rich spring water that attracts fish, particularly salmon (D. Katzeek, 2013, personal communication). The re-labeling of lakes, streams, creeks, and other water bodies to English names of European entrepreneurs erased the storied connection of Tlingit people with water and its circulation.

Narrative ecologies include a second register of storytelling that resides in Tlingit oral histories. The power of these strong stories lies in their ability to shape-shift, like water, according to historical circumstance and the audience's needs. Tlingit *Haa K̲usteeyí* includes mythic time and, like water, are considered to be great teachers. From a Tlingit perspective, stories 'live' and can be self-sustaining; they reside in people and places and perform and generate medicinal guidance for all aspects of the human condition: physical, spiritual, emotional, and intellectual. An excellent example of this is Tlingit Elder Elizabeth Nyman *Gágiwdul àt* (Yanyeidí Clan, Wolf Moiety, Taku River Tlingit First Nation) who also chose to perform at the International Storytelling Festival in 1994. Elizabeth Nyman spoke of an intimate and animate reciprocity between her Yanyeidí Clan and the Taku River. She spoke not only of Yanyeidí Clan histories and claims to the Taku River, but more importantly stressed passing on a legacy from her ancestors to her descendants where the Taku River would be respected. Synthesized by one sentence in her book she was writing at the time of the Storytelling Festival, Elizabeth confirms a deep link with water and stresses a sense of place unparalleled in western science: "If only you were taken by boat along the Taku River, you could write down the whole story in a book" (Nyman in Cruikshank 1998, 149–151).

CTFN Elder Angela Sidney/*Stóow* (Deisheetaan Clan, Crow Moiety) was the older sister of Dora Wedge/*Yàjindahên* (Deisheetaan Clan, Crow Moiety), Mark Wedge's mother. Like Colleen James' grandmother Lucy Wren/*G̲ooch Tláa*

(Dak̲l'aweidí Clan, Wolf Moiety), Angela was one of many Elders to promote and sustain Tlingit oral traditions. She was quite clear that she had "no money to leave to her children – my stories are my wealth" (Angela Sidney in Cruikshank 1990, 36). In collaboration with Julie Cruikshank, Angela Sidney showed how "a single story, carefully told, can be employed to convey different meanings to different audiences." Angela further illustrated that, "[w]hen potential for conflict emerges among people with different perspectives, successful resolution often involves demonstrating how a story can reframe a divisive issue by providing a broader context for evaluating such issues" (Cruikshank 1998, xv).

Water, the stories generated by water/*héen*, and water bodies are all agents in the co-construction of a Tlingit sense of place. Indeed the CTFN Elders' statement at the beginning of the CTFN Final Agreement with the Canadian Government (2005) states very clearly, "We who are Tagish and we who are Tlingit, our heritage has grown roots in the earth since the olden times. Therefore we are part of the earth and the water." Put another way, *héen* has not only kinetic energy, but profound narrative agency. Tlingit storytelling's most animate qualities and its most important living functions are survival and conflict resolution. An excellent example of this is *Héidu dei k̲áa* or "The Man Behind the Dam"—a story about water and the sacred, related in an interview by Mark Wedge (for this story, see the Appendix). While Tlingit stories may indeed chart narrative templates and metaphors for water that offer alternatives to 'modern water,' it is the fluid social life of the stories themselves—their very animistic essence—that is critical to acknowledge. As Robert Bringhurst concurs, "[i]n an oral culture, stories are given voice. They are also given the silence in which to breathe" (2006, 176).

Slow violence and slow activism?

I consider 'slow activism' to be the counter-story to Nixon's 'slow violence,' which he envisions as the gradual, often invisible violence wrought by climate change, toxic drift, deforestation, and oil spills that exacerbates the vulnerability of ecosystems and of people who are poor, disempowered, and often involuntarily displaced (Nixon 2011). I expand the boundaries of slow violence here to include the colonizing of Yukon water through capitalist-driven commodification and the racialized boundaries of modern water. Slow violence has continued through the impacts of the first Whitehorse hydroelectric dam built in 1958, which resulted in a fundamental change in the Southern Lakes water levels. As CTFN Elders describe, the dam created a profound alteration in duck, geese, and swan migratory patterns; muskrat and frog ecologies; and the salmon run, depriving the Southern Lakes and Yukon First Nations of part of their traditional hunting and subsistence foods (Elder Ida Calmegane/*La.óos Tláa*, Deisheetaan Clan, Crow Moiety, 2013, personal communication) (Figure 10.3). CTFN Elder Art Johns/*Ts'eiwát* (Yanyeidí Clan, Wolf Moiety) has spent most of his life on the land and water closely observing these ecological changes: "We don't have any more [musk] rats here because of the fluctuation of the water . . . When that water drains, all the muskrats die off here . . . Everything is gone." He tells of a time when it "used to be just noisy here in the spring with frogs – all over. Now there's none." He agrees that frogs are sensitive environmental indicators. Art calls the frogs 'our canaries.'

Slow violence is also pervasive in water's very invisibility via pipes and the tap, another component of colonized water that privileges the anthropocentric logics of efficiency, profit, and progress (Chen et al. 2013; see also Kambic, this volume). Several Elders talked about 'packing water' (carrying water from the stream to the home) as a practice which not only traces a particular relationship with water, but when completed was often rewarded with the gift of a story by Elders (Elder Winnie Atlin/*Dagé*, Ishkahittaan Clan, Crow Moiety, 2013, personal communication). Colleen James succinctly summarizes why the chore of 'packing water' was critical in the maintenance and evolution of this relationship with water:

> The reason I mention water packing, is because it was a daily thing that we had to do as children. When you walk down to the river, every day you see something there. And the migratory birds come through here all the time. So, in the spring it was really special, because you have swans and geese and everything just right there at the river's edge. The river ice kinda ends a little ways out from the ice hole, so we got to see a lot. Every day as we had to go down to the river, we got to learn what the ice was doing and when it was doing it. When it was safe and when it wasn't . . . so when we were growing up, the water packing was a huge part of our everyday thing. I remember it well but I didn't always like it. But the experiences gained through doing

Figure 10.3 Elder Ida Calmegane/*La.óos Tláa* (Deisheetaan Clan, Crow Moiety) at home in Tagish, August 2013

Photo credit: Eleanor Hayman

Figure 10.4 Elder Norman James/*K̲aakligé* (Dak̲l'aweidí Clan, Wolf Moiety) at Conrad, Windy Arm, Southern Yukon Lakes, August 16, 2013

Photo credit: Eleanor Hayman

> it – the ducks and the swans and the Northern Lights, stars and moon . . . It's something my grandchildren are really disconnected from today.

The disconnect Colleen mentions is deepened by Elder Norman James' reference to distant cultures of consumption through the promotion of bottled water (Figure 10.4). He connects this at a profound level with his Tlingit identity:

> Long time ago this water you could drink any place, any creeks here – good water. I don't know what's wrong with it now – you have to drink your own bottled water. If we start drinking that [bottled] water, I think we're starting to forget who we are.

Whether or not this can be read as part of a post-colonial project of assimilation or the power of the capitalist monoculture remains to be seen. Norman explains how water-related practices are intimately bound up in and stitched into culture. He describes how storytelling and cultural water practices are being eroded by aggressive capitalist forces in the form of 'modern water.' Yet, at a more intimate level, the water humans drink becomes the makeup of the human body and a reflection of ecological relationships.

'Slow activism' centers on the enduring performances of Tlingit storytelling that are bound up in Tlingit oral traditions and the verb-oriented Tlingit language. 'Slow' reflects an underlying resilience and adaptability (not unlike the core characteristics of water) within Tlingit oral tradition that does not pay homage to the western capitalist anthropocentric logic of time=money. Slow activism through storytelling connects to a far older set of philosophies and relationships where qualities such as respect and reciprocity are privileged. For example, there are many *Haa K̲usteeyí* that concern marriages between humans and other animals in order for humans to understand alternative worldviews (e.g. The Girl who Married a Bear). Ten thousand years of storytelling evolved a narrative tradition that has witnessed countless climate changes and has adapted accordingly (see Cruikshank 2005). At a time when many First Nations are struggling to retain identity and coherence in a rapidly changing world, it is the power of strong stories that offers that unique combination of knowledges for conflict resolution and survival.

Conclusion

By way of conclusion and to come full circle, I return to the clarion call for a 'revolution of the mind' and a 'new *water* consciousness' by quoting Baird Callicott (2012), "Ultimately the only effective way to change what we do in regard to the non-human environment is to change the way we think about it."[15] This collaborative water research explores Tlingit relationships with water and water bodies in order to develop legislation rooted in indigenous philosophy for a CTFN Government Water Act. However, it goes further than that by offering important ways to rethink dominant relationships to and understandings of water and gender, and how so often there is the unquestioned assumption that water 'needs' to be managed and that 'gender,' 'feminism,' and 'activism' are fixed categories. These entrenched systems of thought and consequential social and water governance systems are fundamentally challenged by a Tlingit cosmology where humans manage themselves in relation to seasonal fluctuations and the hydrological rhythms of water. For example, two of the most well-known Tlingit narratives describe how *Yéil* (Raven/Crow)—the trickster—contrives sophisticated and ingenious ways to release the sun, moon, stars, and water from the command and control of a single individual (for this story, see the Appendix). Tlingit narratives emphasize a certain form of social justice and freedom from a single, dominating mode of governance (D. Katzeek, 2014, personal communication). This study also calls for a broadening of feminist political ecology frames to include ethnicity.

Alternative narrative templates and metaphors for water are necessary to counter the dominant rhetoric of global water scarcity with its implied water governance generated through western anthropocentric understandings of water, gender, property, and even history. Tlingit storytelling and narratives of water are resilient because they are constantly evolving and shape-shifting. It is these fluid Tlingit narrative ecologies that continue to circulate and thereby transform and inform relationships with water. Might not this slow activism through storytelling water be one of many voices to counter modern water's slow violence?

Acknowledgments

This chapter acknowledges the rich and priceless collaboration of CTFN community consultants Colleen James/*G̲ooch Tláa* and Mark Wedge/*Aan Gooshú*.

My deep thanks go equally to CTFN Heritage Manager, Heather Jones for being such a great facilitator, support, and guide.

To CTFN Elders, community citizens, and staff interviewed and consulted, I thank and honor you, especially CTFN *K̲áa Shaadé Héni* (Chief) Danny Cresswell and the CTFN Government, in particular the Land Management Team.

For Tlingit language guidance I thank Lance Twitchell/*X̲'unei* and David Katzeek/*Kingeistí*.

For providing valuable comments I thank Julie Cruikshank, Heather Sullivan, Lance Twitchell, David Katzeek, David Neufeld, Sheralin MacGregor, Stephanie Buechler, and Anne-Marie Hanson.

This research is supported by the Heinrich Böll Stiftung, Germany, with previous support from LMU, Munich, Germany, and a Northern Research Endowment Grant from the Northern Research Institute, Yukon College.

Notes

1 In the Tlingit language *Héen Kas'él'ti X̲oo* means 'among the ragged lakes' (Sidney 1986).

2 Drawing on Jules Law's introduction and title, *The social life of fluids* (2010), and the philosophy behind Julie Cruikshank's *The social life of stories* (1998).

3 Source: http://penn.museum/collections/shotridge/the_tlingit.html. Accessed February, 2014.

4 More information on the Tagish language with audio examples can be found here: www.ynlc.ca/languages/tg/tg.html. Accessed March 10, 2014.

5 For a comprehensive background to Final Agreements in Canada see the Report on the Royal Commission on Aboriginal Peoples: www.aadnc-aandc.gc.ca/eng/1100100014597/1100100014637. Accessed August 4, 2014.

6 All quotes from Colleen James are taken from personal communication in 2013 and 2014.

7 I draw on but am not limited to Linton's concept of 'modern water' as "intellectual abstraction, scientific specification, material containment and alienation from society and from the rest of non-human nature" (2010, 19). No notion of gendered water is included.

8 I use Margarat Kohn's (2014) reading of the 'colonial project.'

9 See the 44-page strategy here: http://yukonwater.ca/news/updates/2014/06/11/yukon-water-strategy-released! Accessed August 8, 2014.

10 See www.legassembly.gov.yk.ca/pdf/rbhf_ctfn_submission.pdf. Accessed December 4, 2014. "Carcross Yukon crowd say 'no' to fracking." Video produced to record the visit of Yukon's Select Committee on the risks and benefits of hydraulic fracturing to Carcross Yukon on September 24, 2014. For link to vimeo see http://vimeo.com/107123650. Accessed December 1, 2014.

11 See Carcross/Tagish First Nation bans controversial fracking practices in Traditional Territory, PR Newswire, December 5, 2013: www.newswire.ca/en/story/1276597/carcross-tagish-first-nation-bans-controversial-fracking-practices-in-traditional-territory. Accessed February 22, 2014.

12 *Spiritual Connections and Obligations: The Foundation of Tlingit Existence*, Lecture, Steve Langdon, Sealaska Institute, Juneau, Alaska, November 5, 2013.

13 The Tlingit year followed lunar cycles.

14 Elder David Katzeek suggested a word other than 'accounts' or 'stories' which was taken up by Colleen James and CTFN Elders who referenced the Tlingit *Haa K̲usteeyí*.

15 Interview with Baird Callicott undertaken by the Global Environmental Team of the Social and Human Sciences Sector. Interview available at: http://waterculture.us1.list-manage.com/track/click?u=31982f6e4937945bfaddf6712&id=3c58726ed2&e=d83fa9bb9c. Accessed August 8, 2014.

References

Agrawal, A. 1995. Dismantling the divide between indigenous and scientific knowledge. *Development and Change* 26: 413–439.

Benkert, B. and Johnson, L. 2011. *Hydrology of the Bennett Lake Watershed: Contemporary conditions and potential impacts of climate change*. Yukon College, Whitehorse, Yukon Territory, Canada: Northern Climate Exchange.

Bavington, D., B. Grzetic, and B. Neis. 2004. The feminist political ecology of fishing down: Reflections from Newfoundland and Labrador. *Studies in Political Economy* 73 (Spring/Summer): 159–182.

Blaser, M. 2010. *Storytelling globalisation: From the Chaco and beyond*. Durham, NC: Duke University Press.

Bringhurst, R. 2006. *The tree of meaning: Language, mind and ecology*. Berkeley, CA: Counterpoint Press.

Chen, C., J. Macleod, and A. Neimanis, eds. 2013. *Thinking with water*. Montreal, Canada: McGill-Queen's University Press.

Coates, K. 2008. *The Indian Act and the future of aboriginal governance in Canada*. Research Paper for the National Centre for First Nations Governance. http://fngovernance.org/ncfng_research/coates.pdf. Accessed July 1, 2014.

Code, L. 2006. *Ecological thinking: The politics of epistemic location*. Oxford: Oxford University Press.

Cruikshank, J. 1990. *Life lived like a story: Life stories of three Yukon Native elders*. In collaboration with A. Sidney, K. Smith, and A. Ned. Lincoln, NE: University of Nebraska Press.

——. 1998. *The social life of stories: Narrative and knowledge in the Yukon Territory*. Lincoln, NE: University of Nebraska Press.

——. 2005. *Do glaciers listen? Local knowledge, colonial encounters and social imagination*. Vancouver: University of British Columbia Press.

Eigenbrod, R. 2014. Including indigenous knowledges or the challenges of finding an "ethical space." Presentation at annual conference of Gesellschaft für Kanada-Studien (GKS), Grainau, Germany.

Elmhirst, R. 2011. Introducing new feminist political ecologies. *Geoforum* 42: 129–132.

Fox, D. 2014. Antarctica and the Arctic: A polar primer for the new great game. *The Christian Science Monitor*, January 12. www.csmonitor.com/World/Global-Issues/2014/0112/Antarctica-and-the-Arctic-A-polar-primer-for-the-new-great-game. Accessed January 16, 2014.

Gaard, G. 2001. Women, water, energy: An ecofeminist approach. *Organisation & Environment* 14(2): 157–172.

Grey, S. 2004. *Decolonising feminism: Aboriginal women and the global 'sisterhood'* in enweyin, Volume VIII, 2003–2004. www.academia.edu/1330316/Decolonising_Feminism_Aboriginal_Women_and_the_Global_Sisterhood. Accessed May 10, 2014.

Hayman, E. 2012. Shaped by the imagination: Myths of water, women and purity. In *On water: Perceptions, politics, perils*, eds A. Kneitz and M. Landry, 23–33. RCC Perspectives 2012/2. Munich, Germany: Rachel Carson Center for Environment and Society.

www.environmentandsociety.org/perspectives/2012/2/article/shaped-imagination-myths-water-women-and-purity. Accessed January 13, 2014.

——. forthcoming. Whose anthropocene is it? Decolonising the anthropocene in rivers of the anthropocene. In *Rivers of the anthropocene*. http://rivers.iupui.edu/cms/. Accessed August 15, 2014.

Howitt, R. and S. Suchet-Pearson. 2006. Changing country, telling stories: Research ethics, methods and empowerment in working with Aboriginal women. In *Fluid bonds: Views on gender and water*, ed. K. Lahiri-Dutt, 48–63. Kolkata: STREE.

Kohn, M. 2014. Colonialism. In *The Stanford Encyclopedia of Philosophy* (Spring 2014 Edition), ed. E.N. Zalta. http:/plato.stanford.edu/archives/spr2014/entries/colonialism/. Accessed June 15, 2014.

Kovach, M. 2009. *Indigenous methodologies: Characteristics, conversations, and contexts*. Toronto: University of Toronto Press.

Law, J. 2010. *The social life of fluids. Blood, milk and water in the Victorian novel.* Ithaca, NY: Cornell University Press.

Leff, E. 2012. Political ecology – A Latin American perspective. In *Culture, Civilization and Human Society*, [Eds UNESCO-EOLSS Joint Committee], in *Encyclopedia of Life Support Systems* (EOLSS), Developed under the Auspices of the UNESCO, EOLSS Publishers, Oxford, UK. www.eolss.net. Accessed October 30, 2014.

Linton, J. 2010. *What is water? The history of a modern abstraction.* Vancouver: University of British Columbia Press.

——. 2014. Modern water and its discontents: A history of hydrosocial renewal. *Wiley Interdisciplinary Reviews: Water* 1(1): 111–120.

McLeod, N. 2009. Cree poetic discourse. In *Across cultures, across borders: Canadian Aboriginal and Native American literatures*, eds P. DePasquale, R. Eigenbrod, and E. LaRocque, 109–122. Peterborough, Ontario: Broadview Press.

Mignolo, W. 2013. *On pluriversality*. http://waltermignolo.com/on-pluriversality/. Accessed May 12, 2014.

Mollett, S. and C. Faria. 2013. Messing with gender in feminist political ecology. *Geoforum* 45: 116–125.

Nixon, R. 2011. *Slow violence and the environmentalism of the poor.* Cambridge, MA: Harvard University Press.

Sawyer, S. and A. Agrawal. 2000. Environmental orientalisms. *Cultural Critique* 45: 71–108.

Semmens, K., J. Ramage, A. Bartsch, and G. Liston. 2013. Early snowmelt events: Detection, distribution, and significance in a major sub-arctic watershed. *Environmental Research Letters*, 8(1): 11 pp. Open access 014020. http://iopscience.iop.org/1748-9326/8/1/014020/article. Accessed May 27, 2014.

Sidney, A. with J. Cruikshank. 1986. *Place names of the Tagish region.* Whitehorse, Yukon: Yukon Native Languages Project.

Smith, A. 2011. Against the law: Indigenous feminism and the nation-state. *Affinities* 5(1): 1. http://affinitiesjournal.org/index.php/affinities/article/view/73/233#_edn1. Accessed May 10, 2014.

Smith, L.T. 1999. *Decolonising methodologies: Research and indigenous peoples*. Wellington, New Zealand: University of Otago Press.

Suzack, C., S.M. Huhndor, J. Perreault, and J. Barman, eds. 2010. *Indigenous women and feminism: Politics, activism, culture*. Vancouver: University of British Columbia Press.

Turk, A.G. 2011. Exploring philosophy of place: Potential for synergy between phenomenology and ethnophysiography. In *Landscape in language: Transdisciplinary perspectives*, eds D.M. Mark, A.G. Turk, N. Burenhult, and D. Stea, 47–72. Amsterdam: John Benjamins Publishing Company.

11 Pamiri women and the melting glaciers of Tajikistan

A visual knowledge exchange for improved environmental governance

Citt Williams and Ivan Golovnev

Understanding and adapting to climate change requires recognition of the diversity of knowledge sources. At the local level, community perspectives, particularly of indigenous peoples, are largely treated as anecdotal data and outside of current scientific paradigms, even though they represent critical alternative knowledge systems based on many years of observation. In this chapter, we argue from a feminist political ecology perspective that the Euro-Western climate science-to-policy paradigm oppresses alternative worldviews and paralyses public agency through elitist mechanistic science and market-controlled media networks that rob populations of alternative narratives to those of globalized, capitalist-based explanation. By focusing on a visual ethno-climatology case study developed with our research partners in Tajikistan's Pamir Mountains, we contrast two climate knowledge system outcomes and highlight the ongoing marginalization of audio-visual findings within so-called global environmental assessments and communications. By reconfiguring audio-visual methods into global climate assessments, alternative knowledge systems can more equitably enter the environmental change discourse. Through triangulation, visually evidenced and culturally understandable narratives of climate change move us towards improving individual agency within multi-scale climate governance frameworks.

Our study focused on interviews conducted for a video project and outlines a knowledge creation process beyond participatory video to active collaboration video. Our project was a coming together of visual ethnographers from the United Nations University, natural scientists from Tajikistan's national and regional institutions, and rural women as storytellers to co-produce new forms of audio-visual knowledge. We first outline the political development of positivist climate science assessment and its links to dominant policy paradigms. Second, we highlight how climate governance levers are linked to the globally homogeneous climate communications 'products.' We then contrast this with the political emergence of grassroots and indigenous environmental groups and the strategic use of digital audio-visual tools to resist oppressive dominant discourses and to assert self-determination through alternative worldview media narratives. We conclude that our indigenous partners have additional localized indicators of environmental change that are missing from current national and international assessments, indicating the need for reconfiguration of the current unethical and systemic disconnect between climate science assessments, climate change communications, and traditional knowledge systems.

Climate knowledge sources and governance

Contemporary climate science and its terminology are accessed through privileged and technical education, ultimately rendering a non-expert public's reasoning dependant on superior scientific expertise (Heymann 2010). Described as a 'Deficit Model,' the public is often deemed an empty vessel for filling with science expertise to prove 'one truth and one science' (ibid.). Some argue that this campaign is an aggressive colonization of cultural understanding, where a continued dependence on modernity's expert institutes (Nerlich et al. 2009) limits the public's access to visioning and imaging of a self-determined future (Hulme 2011).

Today, inter-governmental climate negotiations managed by the United Nations Framework Convention on Climate Change (UNFCCC) are informed by scientific peer-reviewed assessment outputs of the Intergovernmental Panel on Climate Change (IPCC). Since 1990, the IPCC, a global network of over 831 scientific experts in various environment-related fields, has evolved initial scientific assessments from physical science reviews to multi-disciplinary categorizations of climate change. Its quantitative findings elevate climate change to high priority on the global political agenda. Despite winning a Nobel Prize in 2007, the IPCC has been heavily criticized by scholars in the fields of Political Science, Climatology, Science and Technology, and Geography (Beck 2011; Hulme 2009, 2011; Jasanoff 2010; Radcliffe et al. 2010; Terry 2009) for being an elitist, patriarchal, homogeneous group that produces only positivist, consensus-based science which detaches from local values and "gives decision makers a false sense of confidence in the nature of the problem and solutions needed" (Ford et al. 2012, 212).

Scientific climate knowledge and women

In the past, the female voice in climate change discourse or climate negotiations has been minimal. The IPCC has a poor track record of female membership (Ho-Lem et al. 2011), only recently increasing the proportion of contributing female authors to 25 percent (IPCC 2010). Similarly, the UNFCCC only began addressing female participation in the UN climate negotiations in 2009. With only 30 percent of female delegates present across 2008–2013, critics suggest that in order for a truly global diplomacy, the UNFCCC needs a "50 percent increase in women leaders before international climate agreements really reflect a female perspective" (Petherick 2013, 96).

Despite minimal inclusion of women in the international political negotiation of climate change, in many agrarian regions of the world, women are important actors in land use, water and energy consumption, and fundamentally contribute to the reproductive and productive functioning of the household and community (Agarwal 2010; Dankelman 2010). Extensive reviews of literature exploring gender-related climate change studies indicate that across cultures, women experience the impact of climate change differently, and have different responses and support needs than men (Aboud 2011; Dankelman 2010). Abrupt climate-related shifts threaten to increase existing gender inequities. Moreover, current climate assessment and policy frameworks created to address the inequities often ironically reflect and reinforce both gender and regional inequalities (Ho-Lem et al. 2011; Petherick 2013; Root 1993; Terry 2009). In particular, weak or excluded political representation further renders women vulnerable to the existing inequities in resource and information access, socio-economic status, and exacerbates direct needs connected to physical and mental health, food security, and child-rearing (Aboud 2011; Ho-Lem et al. 2011; Petherick 2013). Furthermore, the lack of sufficient female voices in climate negotiations ultimately deprives society of the possibility of a diverse council for improved and inclusive environmental governance.

Climate media

Historically, many individual societies made sense of the world through the artistic storytelling of archetype myths and narratives conveyed around a central performance point such as campfire, heath, or stage (Campbell 2008). Through ritualized performance, members of a socialized group engaged and collectively enforced social order through role models, cultural practices, and norms (Johansson 2002). Today, Western society's mass media still communicates within ritualized storytelling environments. Through culturally embedded meanings (ibid.) media plays to an audience's cultural cognition, creating an artistic framing of messages that give meaning or reaffirm a social group's cultural values (Kahan et al. 2011). Mass media portrayals of climate change can be a powerful mediator of reality cognition. In a broad sense, if the framed narrative does not resonate or evolve meaning within the viewer's long-term memory, and 'deep framings' of identity or experience (Breakwell 2010; Crompton 2010), then its message is often ineffective, potentially even polarizing the viewer's belief away from the intended message (Kahan 2010).

Studies indicate that the non-expert public in Western societies gather most knowledge of environmental change from the mass media (Boykoff and Smith 2010; Wilson 1995). European and North American societies often rely on consumer-skewed network media, dominated by discourses of Western science, enlightenment age classification, and economic risk rationality. Furthermore, the majority of climate change mass media solutions are framed as market transitions, avoiding the moral inequities of the social-environmental relationship while continuing to censor or confuse cognitive attempts to frame life beyond capitalism (Bowman 2010). As a stage where only elite and salient actors have access to debate climate governance (ibid.), the use of well-crafted propaganda reinforces deeply framed cultural norms and directs populations towards powerful unseen financial portfolio interests that are not necessarily in the interests of the civilian population (Altheide et al. 2001). This type of mass media also restricts public access to alternative knowledges of climate change (Heyd and Brooks 2009; Himelfarb et al. 2009).

Climate communications research within various regions of Africa, Asia, and India indicates links to European and North American media's use of the scientific deficit model to communicate climate change (Boykoff 2010; Boykoff and Smith 2010). Yet climate change is a polycentric concept, meaning different things to different people living in diverse geographic regions. For example, a BBC media study across several African countries reported that publics were confused by the term 'global warming' and how it connected to previous ozone hole campaigns (Godfrey et al. 2012). It is from within the margins of this cultural imperialist media landscape that indigenous media movements are now emerging (Alia 2010).

Indigenous and community knowledge and climate

Many indigenous peoples live at the frontline of anthropogenic climate change due to resource-dependant life-ways, territories ranging across marginalized ecosystems, and wellbeing sensitively connected with nature and its cycles (Green and Raygorodetsky 2010; Nakashima et al. 2012). Recognized by the United Nations as unique members of the global community, indigenous peoples manage or utilize territories that harbor 80 percent of the world's remaining biodiversity (Nakashima et al. 2012). Nevertheless, through factors including complex multi-generational discrimination, lack of access to resources, and the compound effects of linguistically insensitive educational systems, indigenous peoples constitute 15 percent of those living in extreme poverty around the world (UNPFII 2009). With largely oral-based cultures and some of the lowest literacy rates in the world (Hanemann 2004), cultural cognition of mainstream climate change does not reflect or represent many indigenous cultural configurations. Within ongoing colonization, indigenous peoples struggle to steward differing worldviews, oral traditions, and long-term place-based traditional knowledge across successive generations (Berkes and Turner 2006; Cruikshank 2001; Turner and Lantz 2003; UNPFII 2009). The IPCC Assessment Report 4 in 2007 recognized indigenous knowledge as an invaluable source (Parry et al. 2007), but referenced mainly Arctic research (Ford et al. 2012). Assessments by indigenous peoples themselves were mostly epistemologically classified as grey literature and deemed inaccessible by peer review due to socio-cultural barriers, though a small cross-cultural research network is making important headway (ibid.; Galloway McLean et al. 2012; Nakashima et al. 2012).

Over the last 40 years, indigenous peoples' demands for land rights and self-representation have led to a critical explosion of indigenous media. Many low literacy community groups are accessing digital technologies to suit culturally appropriate frameworks for conveying news, documenting events, combating discrimination, advocating rights, and carrying out cultural maintenance activities (Alia 2010; Banks 2001; Janke 2009; Wilson and Stewart 2008). While critics label the movement as neo-colonialism, in which the media elite assimilate indigenous identity through technological determinism (Ravetz and Sardar 1996; Sardar 1995; Stam and Shohat 1994), many indigenous groups consider these media outlets as adaptive measures for cultural survival and solidarity (Ginsburg 2008; Michaels 1986). Digitally networked stories play an important role in the emancipation of marginalized knowledge systems into the digital realm (Alia 2010; Ginsburg 2008; Soukup 2006; Srinivasan 2012). Indigenous peak bodies and international development agencies support indigenous media projects for political strategizing and the transmission of social values, knowledge, and human resource development (UNEP 1992; Wilson and Stewart 2008). Indigenous audio-visual work on climate change is also emerging, but its intended knowledge-diversifying role in global climate assessments is thus far undervalued and largely considered as anecdotal or advocacy communications. Our collaborative media project arose from the necessity to address the issue of taking seriously indigenous women's knowledge of climate change as it relates to both local and global socio-environmental change and the melting glaciers in the Pamir Mountains of Tajikistan.

Research site

Our case study location of the Gorno-Badakhshan Autonomous Oblast (GBAO) region of Tajikistan is a bio-culturally diverse mountainous landscape with complicated terrain and continental climatic variability proving challenging for climate modelers. It is geographically characterized by high 5,000–7,000-meter mountain ranges separated by steep river valleys situated 1,700–2,500 meters above sea level. The region is approximately 25,700 km^2, but its sparse population (7.6 persons per km^2) still puts pressure on resources available due to the short summer growing season and steep mountain desert terrain yielding 0.04 percent arable land (Nabhan et al. 2012; PALM Project 2011). Most Western Pamir villages employ livestock-raising and mixed mountain agriculture, which relies on stream-fed irrigation heavily influenced by melting snow and glaciers during the summer (PALM Project 2011). Home to a number of rare and endemic flora, its ancient petroglyphs indicate that rare wildlife species such as Siberian Ibex and Snow Leopard have lived among the Pamiri mountain folk for many generations (ibid.).

National climate assessments and projections based on satellite data, modeling, and sparse instrumental records spanning back to the 1930s suggest that by 2030 Tajikistan will experience an increase in mean annual temperature of 0.2–0.4°C (Makhmadaliev et al. 2008). German-developed global atmospheric/ocean circulation model ECHAM4/OPYC3 indicates a likely increase in Western Pamir annual precipitation, largely due to prevailing westerly Jetstream conditions, spatial precipitation fluctuation, and high altitude Western facing topography (Kayumov and Rajabov 2010). Changes in temperature and precipitation are expected to expose Pamiri people to extreme weather events and increased precipitation variability triggering avalanches and mud falls, heavy rainfalls, increasing wind speeds and direction, and glacial lake 'flash flooding' outbursts (ibid.; Makhmadaliev et al. 2008).

Instrumental records and composite satellite images indicate that over the last 50 years there has been a reduction in glaciation area by 25–30 percent. These glaciers play an important role in the Amundarya River (the largest artery of water for Central Asia), with impacts on glacial volume having a critical effect on downstream populations in arid areas across Central Asia (Kayumov and Rajabov 2010). In the short term, while wheat and fruit growing in high altitudes is seen as positive for mountain villagers, localized discussions among local and regional leaders may benefit from the dovetailing of climate modeling, adaptation strategies, and established community resilience measures (Kassam 2009).

Gender relations and audio-visual ethnography in the Pamir region

Gender relations in the Pamir region are influenced by a number of historical and cultural traditions, ethnic, religious, and geographical features. An increasing interest in and relevance of gender is due to the increasingly public role of women during the post-Soviet times, Tajikistan's economic growth, and academic and gender policy focus in social and political activities of Tajikistan (Bliss 2005; Nabieva 1999). During the Soviet period, there were powerful state interventions in family life designed to assist women in career growth and economic productivity (Karimova 1980). More recently, political and economic changes, together with the globalized influences of mass media and culture, have caused some traditional values (e.g. family models, marriage customs, women's family roles) to be rejected on the basis that traditional or indigenous values promote a negative effect in modern Tajikistan society (Zakrieva 2001).

While in other regions of the country the social passivity of women is cultivated, in the Pamir region, gender relations are generally determined by traditional cultural attitudes towards mountain women as always having high social status and relative freedom (Bliss 2005). The ongoing perspectives and increasing role of women in modern Pamir society are important because of issues affecting local social stability including: opium drug addiction among mountain populations, cultural marginalization and corrosion, unemployment, migrant movements, and economic uncertainty (ibid.).

Ethnographic video as process

Visual ethnographic methods were focused on the Pamir region as early as 1928. In those days, the scientific gaze surveyed the exotic other (Erofeyev 1928). Regional visual ethnographic studies across the twentieth century did not specifically focus on gender, although women were present constantly as active caretakers and mothers. Post-Soviet depictions reflect women in a dynamically transforming culture involved in both public and private sphere (e.g. see works "Two" or "Where the sky" from Tajik filmmaker Gulbakhor Mirzoyeva). The use of visual methods within mixed methods approaches in part responds to the call by Rocheleau (1995) and Nightingale (2003) to employ mixed methods to address gendered subjectivities in relation to environmental change. Elmhirst (2011) stresses that new feminist political ecologies should include multi-dimensional aspects of subjectivities such as ethnicity and encourages a multi-scalar examination of power and representation in decision-making.

This project began in 2009, when in our capacity as qualitative visual researchers we were involved in a small United Nations University (UNU) project, supporting the voices of indigenous peoples in inter-governmental climate negotiations. During a brief meeting in Palo Alto in 2008, a small ad-hoc climate change working group of the United Nations Permanent Forum on Indigenous Issues (UNPFII) met to discuss strategic engagement of indigenous perspectives in international climate discourse. At this meeting it was decided to carry out various grassroots indigenous-led climate assessments using methodologies (IPCCA 2013) that felt familiar and more appropriate to traditional peoples, including audio-visual storytelling (Chilisa 2011).

Previous to this Palo Alto meeting, our small audio-visual team at the UNU in Tokyo had been monitoring the increasing use of audio-visual social media communications among the international relations sector of policy advisors, think tanks, and government departments. We had begun developing the 'video brief,' a short six-minute shareable audio-visual work specific for policy negotiation, public communications, and post-graduate educational environments. The format was an idea adapted from the text-based 'policy brief,' a well-used international relations intelligence format. The video brief became a short and succinct multi-lingual audio-visual report authored by collaborators and supported by us as creative technicians, to be shared largely online through bloggersphere, regional broadcast, UN networks, and in DVD format. Our methodology was influenced by the Central Australian Aboriginal Media Association (CAAMA) and its TV series 'Nganampa Anwernekene' ('ours'), that as a professional Aboriginal media collective adapted visual ethnographic methods to create cultural maintenance works valuing first-person traditional language narratives and ethical community protocol processes.

Snowball sampling brought us to the Pamir case, as the work was commissioned and initiated by the Christensen Fund, a small philanthropic fund based in Palo Alto, California who contributed financial support to the UNPFII efforts. Initially, we had conducted several conversations via telephone and email with our Tajiki field research partners discussing preparation, as well as exchanging various relevant background materials like written reports, videos, letters, and photos.

After arriving in Tajikistan, we met with Professor Abdusattor Saidov from the Institute of Zoology and Parasitology and carried out a semi-structured interview with Professor Kayumov, one of Tajikistan's leading glaciologists working for the Government's State Agency for Hydrometeorology, Committee for Environmental

Protection. During this interview, we were briefed on national climate change assessment evidences, post-Soviet institutional challenges, and the importance to corroborate qualitatively the region's scarcity of datasets. Another partner, regional biodiversity expert Dr Abdulnazar Abdulnazarov, also greeted us upon our arrival at GBAO's capital Khorog. Recognizing the lack of women in positions of public authority, we decided to juxtapose these scientific assessments with cross-generational qualitative materials developed with women, to evidence that differing perspectives and knowledge sets were missing in the national assessments.

Our video was created in three locations across Western Pamir: Khuf Valley, Bulunkul, and Khorog (GBAO's regional capital) (see Figure 11.1). We aimed to facilitate the story the community wanted to tell about environmental change, and worked with suggested storytellers who were comfortable and wished to share a clear view on local lifestyle, observations, and noteworthy changes in the community.

Our group developed a rough story script, location schedule, and shot-list, with a director/camera operator and producer/sound recorder adding technical craftsmanship to the data. We visited and carried out everyday activities at certain sites, filming as we went, and recording semi-structured and then unstructured interviews in traditional language to comfort and capture the candid manner of the storyteller. Each evening after the collecting, the producer, working with the cultural liaison (with others sometimes watching, listening, and clarifying), translated the interviews. These transcripts were discussed, coded manually, and entered into the post-production laptop for editing into an agreed sequence. Other partners included a local female singer, Abrigul Oshurberova, who sang a traditional Pamiri lullaby she felt contributed to the topic and subject matters being discussed. We also worked remotely via phone with Professor Kayumov who helped us to refine Tajikistan's key climate projections texts (Makhmadaliev et al. 2008) to triangulate with the material we had recorded. At the end of the process, once we had carried out various consultation screenings for approval of cultural content and grammar, we made three subtitled versions (Tajik, Russian, and English) and then held three large local Q&A screenings to various targeted audiences including local residents, Western scientists, and NGO communities.

Finally, our research partners were given a small hard drive with the film and raw footage, together with a non-exclusive licence to re-use the materials we had jointly created after agreement by all was reached regarding re-use consents if a third party becomes involved in the re-use. This Intellectual Property dimension of the work was based on indigenous Australian media protocols developed by Janke (2009). We shared the film with local and national broadcasters (though broadcast was never confirmed) and in Tokyo we backed up, polished (coloring, high resolution graphics, and sound mixing), and uploaded our work to YouTube, a robust video delivery system in low bandwidth countries. The work was distributed widely through UN networks, NGO listservs, broadcast/online media networks, and later screened at several film festivals, including at a side event at the UNFCCC COP15 in Copenhagen 2009. The final film is available for viewing at: http://ourworld.unu.edu/en/tajikistan-women.

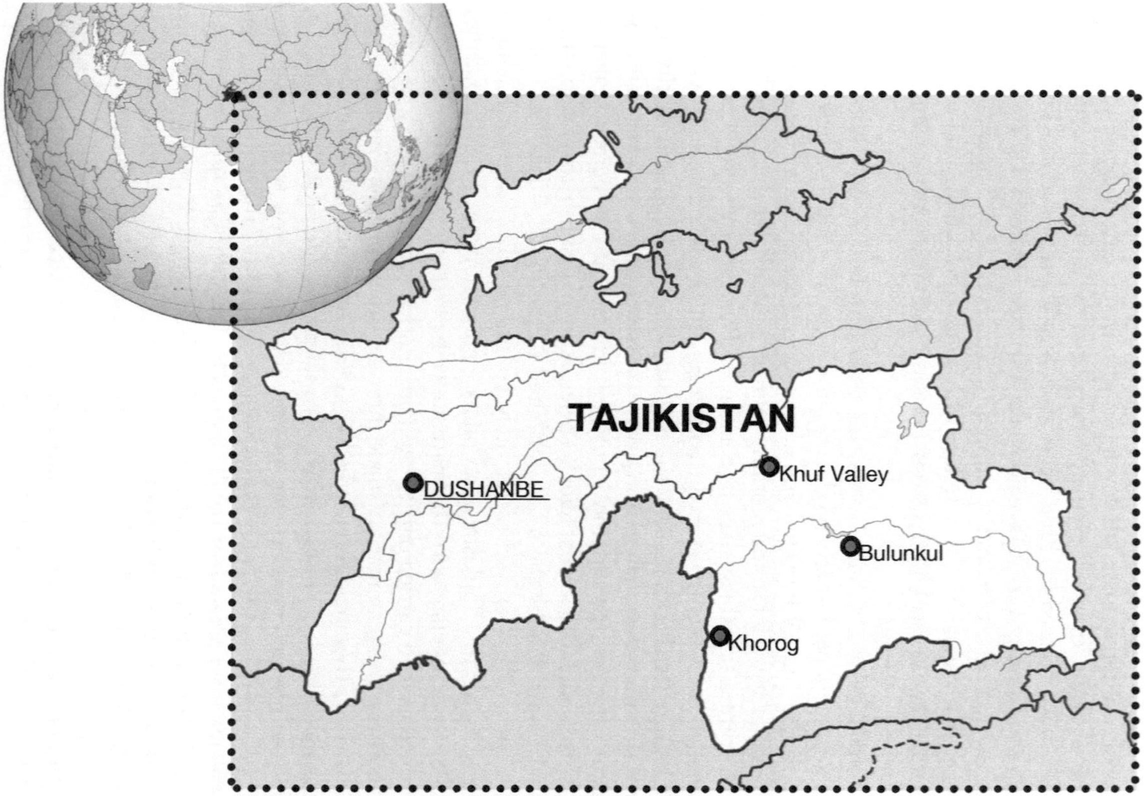

Figure 11.1 Location of case study sites in Eastern Tajikistan

Source: NordNordWest 2008

Pamiri women case study

Through our results, we were able to provide evidence for three Pamiri women's perspectives to changing environmental resources. Traditionally, Pamiri women carry out domestic and light agricultural activities, but are rarely given responsible positions in regional resource governance. However, as harvesters, weavers, millers, herders, gatherers, and providers, women demonstrate keen observation skills and draw upon traditional intergenerational knowledge to adapt within a varying environment. Our partnership documented considerable scope for future collaboration around key issues raised, but also identified a need to take into consideration communication approaches that are culturally respectful. In our interviews, women focused concerns on water and food security, flock numbers, pasture quality and predator appearances, community health (particularly children and extended kin), and environmental changes restricting essential domestic tasks.

In the Khuf Valley, television and media players are present in some houses through micro-hydroelectric power. In this community, 75-year-old Kurbonbegim eagerly wanted to show her life-way and its connection to the glacier at the top of the valley. She invited us to record her everyday life activities at several locations across the valley including harvesting hay, flour grinding at the stream-fed mill, and making the daily bread (where we agreed to interview her in motion). She discussed the reliance on irrigation from the glacial stream for the summer crops like wheat and potato (see Figure 11.2). Remembering her childhood, she also commented that apples and apricots were now growing at higher altitudes in the valley. We later corroborated this with recent agroforestry studies indicating climate envelope shifts taking place in the valley, in particular an upwards elevation shift in crops species of 311m between 1890 and 2006 (Nabhan et al. 2012). As indicated below in excerpts from the digital film transcript, she also mentioned the lack of Ibex (wild mountain goat) sightings and increased wolf appearances in the valley, possibly indicating food scarcity across the higher alpine meadows.

00:02:06,560 Before this, there were more Ibex.
00:02:10,720 Now, you don't see them.
00:02:15,000 This year, there were many wolves that killed a lot of livestock.
00:02:19,440 Apples and mulberries are now growing in our village
00:02:21,800 and fruit trees are now growing in Parjuug (village downstream).
00:02:24,800 Trees don't grow near our place because it is cold.
00:02:28,200 Our place is too close under the mountain,
00:02:32,200 but it will be better in the future.

In Bulunkul, a small village on the banks of a high altitude freshwater glacial lake, we discussed climate change with Bakhitbegim Nebieva, who worked with her husband to keep the local meteorology station. Her presence as meteorological observer offered an unexpected scientific element to the

Figure 11.2 Glacial stream in the Pamir Valley, Tajikistan
Photo credit: Damon Lynch, University of Minnesota

narrative. During the interview and at various planned locations she indicated extreme temperature event fluctuations in the valley and its impact on supply routes and community health (see Figures 11.2 and 11.3). She also discussed the multi-decadal evapotranspiration of nearby glacial lakes and its impact on village food security, health, general livestock numbers, and migratory bird patterns.

00:05:04,720 When it's minus fifty or sixty degrees, it's difficult for our livestock.
00:05:10,680 The weather is unstable, sometimes cold sometimes warm.
00:05:14,320 This is not good for our health.
00:05:17,640 Because of this year's extreme weather, all the people were sick.
00:05:23,320 All people have Rheumatism,
00:05:25,280 sore legs, sore hands.
00:05:33,120 The lake is shrinking . . .
00:05:34,840 smaller and smaller.
00:05:37,480 Bulunkul is disappearing,
00:05:40,720 it's evaporating.
00:05:41,720 It is having a very strong impact.
00:05:45,120 There is not enough grass for our livestock.

In the city of Khorog, we worked with a young 17-year-old woman, Madina Iftikhoz. Her perspectives on environmental change indicated knowledge of

Figure 11.3 Women walking along agricultural fields in rural area in the Pamir Valley, Tajikistan

Photo credit: Damon Lynch, University of Minnesota

pollution and its relationship to the transport sector. Through the knowledge and aspirations communicated by the youngest woman, we found evidence of urbanization eroding traditional bio-cultural relations. In particular, the young Pamiri woman indicated a cognitive disconnection or lack of awareness of localized environmental changes in urban centers. This disconnect was more present here than in the rural areas we visited.

00:08:44,360 There are many cars in the city, it's good for us.
00:08:47,240 It allows us to get around quickly and work faster.
00:08:51,840 On the other hand, more cars are not so good.
00:08:54,120 Cars contaminate the environment and it's bad for our health.
00:09:01,920 My friends and I never talk about it (climate change).
00:09:06,240 We don't know enough about it.
00:09:10,600 Adults don't talk about it either.

Following the collaborative production, the finished film was screened at three separate events. First, we facilitated a screening in the local Khorog town park, where over 80 people were in attendance. The screening gave the community an informal access point to discuss awareness of localized environmental change. Second, we were invited to present a Q&A screening to over 40 Tajik scientists at a Biodiversity conference simultaneously taking place in Khorog. Third, we had a screening and forum discussion with several

Tajikistan Academy of Science experts (all male) and three women-focused NGOs at the Dushanbe offices of the United Nations Development Program (UNDP). During the forum, several male scientific academics rejected the study based on our credentials to speak on the subject, while the women present approached us afterwards with ideas of incorporating the work into community awareness programs. The film was also later screened at the National Museum of Denmark (Copenhagen) as a UNFCCC COP15 2009 five-day side event. Our research partner, Professor Abdusattor Saidov, was in attendance for the lively Q&A session afterwards.

Through our initial fieldwork and video outputs, we were able to technically evidence and enhance localized indicators of environmental change missing from regional and nationally representative climate assessments and communications. These indicators were deemed culturally important to a group of multi-generational women in the remote mountainous GBAO of Tajikistan. Our results indicate disconnects in climate discourse between climate science-to-policy assessments and localized and indigenous knowledge systems surrounding climate change. By developing inclusive and pluralistic climate assessments that value the diversity of its peoples' expertise as sensitive localized functions of observation, monitoring, and intervention, Tajikistan can strengthen its overall adaptive capacity to ongoing environmental change.

Past acknowledgments of inequitable research dynamics have stirred indigenous scholars to articulate new resistant forms of research grounded in "relational accountability promoting respectful recognition, reciprocity and the rights of the researched" (Chilisa 2011, 39–41; Brown and Strega 2005). Dynamic methodological approaches have enabled the development of social theory informed by collaborative inferences from language, oral traditions, ceremony, document storytelling, and artifacts (Chilisa 2011, 39; Gubrium and Harper 2013; MacLean and Woodward 2013; Srinivasan 2007; Wilson 2008).

Within rural communities who have alternative education systems to text, such as many communities in the Pamir Mountain region, inclusive audio-visual methods are accountable tools for data gathering, analysis, and interpretation, enabling triangulation with other sources of knowledge (Chilisa 2011). As outlined, storytelling is a powerful social process, and complex social experiences and internalized cues connected to environmental change are difficult to capture and embed within an individual's cultural cognition through text-based reports or foreign media framing. Collaborative audio-visual methods have the power to de-stabilize instituted power relations affecting indigenous and poor women, who often miss out on literacy education opportunities due to domestic responsibilities or other cultural barriers. Although the cross-referencing of such novel visual data with Western climate science reports is still minimal (Pink 2012), women involved in our video project expressed a sense of emancipation through their participation in climate knowledge production, or as other feminist scholars similarly describe, through the creation of an 'alongside' gaze (Kindon 2003; Pink 2007).

Our study focuses on visual research methods and outlines a knowledge creation process beyond participatory video to active collaboration video. Our work was a coming together of scientific filmmakers from the UNU, scientists from Tajikistan's national and region institutions, and rural women as knowledge holders and storytellers to co-produce new forms of audio-visual knowledge for analyzing, sharing, and teaching. The work outlined a set of guiding principles for collaborative knowledge production and researcher accountability. These principles include: recognition of co-production ownership and full participation; co-creation of visual script and study parameters; non-exclusive licensing of specific cultural materials respecting custodial and intellectual property; first-person community storytelling and language translation; analysis and report editing on location with various community screening approvals; local language versioning; final community consent event; provision of a back-up hard disk copy of all materials created; regular feedbacks about report dissemination; and the support and involvement of local representatives in international climate discussions.

Through facilitated curation across digital networks and screenings, the work has been viewed at inter-governmental side events, academy of science venues, in other remote indigenous villages (e.g. in Peru, Brazil, Indonesia, Australia, Russia), in postgraduate classrooms, and on YouTube over 29,300 times. Since its creation, several local partners have leveraged the experience to forge new partnerships with regional scientific groups shaping guidelines for future media collaborations and culturally inclusive climate toolkits. Most recently, Tajikistan's national correspondence to the UNFCCC includes a gender-based adaptation initiative and a new series of programs supported by the Asian Development Bank. Activities within this initiative include local language gender specific capacity building at community centers through training and equipment (Abt Associates and CLIMsystems 2013). Although a large part of the initiative is thus far based on top-down gender inclusion strategies, our collaborative video work has made an impact on gender and indigenous climate negotiations in the region.

Our project is one example from a small but growing group of dedicated indigenous peoples and environmental scientists who use collaborative audio-visual methodologies in conjunction with satellite imaging and climate modeling in order to incorporate multiple forms of socio-environmental knowledge into the international climate discourse. Although such collaborations are technologically reliant, many indigenous communities are actively engaged with digital media as a form of "strategic traditionalism" (Ginsburg 2008, 19) in order to resist and challenge dominant climate discourses, demand equality and security, and indigenously shape identities and re-territorialize digital depictions of kin, culture, and homelands (Brown and Strega 2005; Dyson and Underwood 2006; Galloway McLean et al. 2012; Mustonen and Mustonen 2010; Tricia Fragnito and Lewis 2005). For example, one video project from Australia's Arnhem Land Plateau explores a 'two toolkit' approach to traditional fire ecology and life-giving waterholes (Garde et al. 2009; Yibarbuk et al. 2001). Another set of studies across the Arctic illustrates how NASA satellite imaging researchers and networked herder groups are respectfully co-producing new multi-scale water regime knowledge (Maynard et al. 2011). Examples placing emphasis on storytelling to access oral histories (Bonny and Berkes 2008; Cruikshank 2001; Johnson 1992; MacLean and Woodward 2013; Russell-Smith et al. 1997) led us to consider audio-video methods as a capable means of generating an environmental knowledge 'boundary object,' described by Zurba and Berkes (2013, 2) as "a focal point for deliberation" and cultural exchange. Our collaborative work in the Pamir region has provided stakeholders with a technically professional 'boundary object' to acknowledge a more respectful communication of Pamir environmental change relating to water regimes, as well as to provide qualitative evidence and recognition of alternative knowledge and value systems already locally in place.

Conclusion

The Western climate science-to-policy paradigm paralyses public agency through elitist mechanistic science, market-driven governance decisions, and globally dominant consumer skewed media network products. This approach to environmental governance is oppressive for peoples with different cultural configurations. The rise of indigenous media from various traditional homelands has brought forth a powerful repurposing of media instruments towards self-determination, dominant discourse resistance, and survival in the face of socio-technical and cultural change. Video projects on climate change and water that give voice to indigenous women can help to focus attention on marginalized knowledge systems and foster inclusion of indigenous women as active negotiators in multi-scale discussions such as the IPCC and the UNFCCC.

We conclude that with respectful collaboration with our local partners, we provided evidence of localized indicators of environmental change missing from current national assessments, indicating a clear disconnect between climate science assessments, communications, and traditional knowledge systems. By reconfiguring global climate assessments to include valid audio-visual methods being explored by various local groups, multiple environmental change knowledge systems and associated parameters can more equitably enter environmental change dialogues. Inclusion is a step towards improving multi-scale climate governance frameworks by including multiple and diverse culturally cognitive narratives that are produced and understood by climate scientists, policy makers, and non-expert local communities. In this regard, we propose further research that includes new forms of collaborative audio-visual 'climate memory' works, multi-media narratives co-coded for rich qualitative climatic data, social memory, and cultural maintenance. Such collaborative work has already impacted regional climate programs in the Pamir and other regions, and has the potential to lead to successful cultural cognition of alternative knowledge systems and globally inclusive multi-cultural climate assessments and communications.

References

Aboud, G. 2011. Gender and climate change: Supporting resource collection. Brighton, UK: BRIDGE, Institute of Development Studies, University of Sussex. www.bridge.ids.ac.uk/go/home&id=59291&type=Document&langID=1. Accessed December 28, 2013.

Abt Associates and CLIMsystems. 2013. TA–8090: Building capacity for climate resilience. Technical assistance consultant's report. Prepared for the government of Tajikistan and the Asian Development Bank. Washington, DC: Climate Investment Funds. www.climateinvestmentfunds.org/cif/sites/climateinvestmentfunds.org/files/ADB_TA_8090_Inception_Report.pdf. Accessed December 26, 2013.

Agarwal, B. 2010. Does women's proportional strength affect their participation? Governing local forests in South Asia. *World Development* 38: 98–112.

Alia, V. 2010. *The new media nation: Indigenous peoples and global communication*. Oxford: Berghahn Books.

Altheide, D., B. Gray, R. Janisch, L. Korbin, R. Maratea, D. Neill, J. Reaves, and F. Van Deman. 2001. News constructions of fear and victim: An exploration through triangulated qualitative document analysis. *Qualitative Inquiry* 7: 304–322.

Banks, M. 2001. *Visual methods in social research*. London: Sage.

Beck, S. 2011. Moving beyond the linear model of expertise? IPCC and the test of adaptation. *Regional Environmental Change* 11: 297–306.

Berkes, F. and N.J. Turner. 2006. Knowledge, learning and the evolution of conservation practice for social-ecological system resilience. *Human Ecology* 34: 479–494.

Bliss, F. 2005. *Social and economic change in the Pamirs (Gorno-Badakhshan, Tajikistan)*. Trans. N. Pacult and S. Guss, with T. Sharp. London: Routledge.

Bonny, E. and F. Berkes. 2008. Communicating traditional environmental knowledge: Addressing the diversity of knowledge, audiences and media types. *Polar Record* 44: 243–253.

Bowman, A. 2010. Are we armed only with peer-reviewed science? The scientization of politics in the radical environmental movement. In *Future ethics: Climate change and apocalyptic imagination*, ed. S. Skrimshire, 173–196. London: Continuum Press.

Boykoff, M. 2010. Indian media representations of climate change in a threatened journalistic ecosystem. *Climatic Change* 99: 17–25.

—— and J. Smith. 2010. Media presentations of climate change. In *Routledge handbook of climate change and society*, ed. C. Lever-Tracy, 210–218. London: Routledge.

Breakwell, G. 2010. Models of risk construction: Some applications to climate change. *Wiley Interdisciplinary Reviews: Climate Change* 1: 857–870.

Brown, L.A. and S. Strega. 2005. *Research as resistance: Critical, indigenous, and anti-oppressive approaches*. Toronto: Canadian Scholars' Press.

Campbell, J. 2008. *The hero with a thousand faces*. Novato, CA: New World Library.

Chilisa, B. 2011. *Indigenous research methodologies*. Thousand Oaks, CA: Sage.

Crompton, T. 2010. Common cause: The case for working with our cultural values. London: WWF-UK. http://assets.wwf.org.uk/downloads/common_cause_report.pdf. Accessed December 28, 2013.

Cruikshank, J. 2001. Glaciers and climate change: Perspectives from oral tradition. *Arctic* 54: 377–393.

Dankelman, I. 2010. *Gender and climate change: An introduction*. London: Routledge.

Dyson, L.E. and J. Underwood. 2006. Indigenous people on the web. *Journal of Theoretical and Applied Electronic Commerce Research* 1: 65–76.

Elmhirst, R. 2011. Introducing new feminist political ecologies. *Geoforum* 42: 129–132.
Erofeyev, V. 1928. *World roof* (in Russian). Dushanbe, Tajikistan: State Archive of the Republic of Tajikistan.
Ford, J.D., W. Vanderbilt, and L. Berrang-Ford. 2012. Authorship in IPCC AR5 and its implications for content: Climate change and indigenous populations in WGII. *Climatic Change* 113: 201–213.
Galloway McLean, K., S. Johnston, and A. Ramos Castillo. 2012. The role of indigenous peoples in global environmental governance: Looking through the lens of climate change. In *Green economy and good governance for sustainable development: Opportunities, promises and concerns*, ed. J.A. Puppim de Oliveira, 245–266. Tokyo: United Nations University Press.
Garde, M., B.D.A. Nadjamerrick, and M. Kolkkiwarra. 2009. The language of fire: Seasonality, resources and landscape burning on the Arnhem Land Plateau. In *Culture, ecology, and economy of fire management in North Australian savannas: Rekindling the wurrk tradition*, eds J. Russell-Smith, P. Whitehead, and P. Cooke, 85–164. Collingwood, Australia: CSIRO.
Ginsburg, F. 2008. Rethinking the digital age. In *The media and social theory*, eds D. Hesmondhalgh and J. Toynbee, 127–144. New York: Routledge.
Godfrey, A., M. Burton, and E. LeRoux-Rutledge. 2012. "Africa talks climate": Comparing audience understandings of climate change in ten African countries. In *The handbook of global media research*, ed. I. Volkmer, 504–520. London: John Wiley and Sons.
Green, D. and G. Raygorodetsky. 2010. Indigenous knowledge of a changing climate. *Climatic Change* 100: 239–242.
Gubrium, A. and K. Harper. 2013. *Participatory visual and digital methods*. Walnut Creek, CA: Left Coast Press.
Hanemann, U. 2004. Literacy for special target groups: Indigenous peoples. Paris: UNESCO. http://unesdoc.unesco.org/images/0014/001460/146004e.pdf. Accessed December 26, 2013.
Heyd, T. and N. Brooks. 2009. Exploring cultural dimensions of adaptation to climate change. In *Adapting to climate change: Thresholds, values, governance*, eds W.N. Adger, I. Lorenzoni, and K.L. O'Brien, 269–282. Cambridge: Cambridge University Press. http://nickbrooks.org/publications/Heyd_Brooks09_Final_proof.pdf. Accessed December 29, 2013.
Heymann, M. 2010. The evolution of climate ideas and knowledge. *Wiley Interdisciplinary Reviews: Climate Change* 1: 581–597.
Himelfarb, S., T. Gould, E. Martin, and T. Sonenshine. 2009. Media as global diplomat. United States Institute of Peace report. www.voltairenet.org/IMG/pdf/Media_as_Global_Diplomat.pdf. Accessed December 28, 2013.
Ho-Lem, C., H. Zerriffi, and M. Kandlikar. 2011. Who participates in the Intergovernmental Panel on Climate Change and why: A quantitative assessment of the national representation of authors in the Intergovernmental Panel on Climate Change. *Global Environmental Change* 21: 1308–1317.
Hulme, M. 2009. *Why we disagree about climate change: Understanding controversy, inaction and opportunity*. Cambridge: Cambridge University Press.
——. 2011. Reducing the future to climate: A story of climate determinism and reductionism. *Osiris* 26: 245–266.
IPCC (Intergovernmental Panel on Climate Change). 2010. 831 Experts Selected for the Fifth Assessment Report. IPCC Press release. www.ipcc.ch/pdf/press-releases/pr-23june2010.pdf. Accessed December 26, 2013.

IPCCA (Indigenous Peoples' Biocultural Climate Change Assessment Initiative). 2013. Methodological toolkit for local assessment. Cusco, Peru: Asociación Andes – IPCCA. http://ipcca.info/toolkits/ipcca-methodological-toolkit.pdf. Accessed December 26, 2013.

Janke, T. 2009. *Pathways & protocols: A filmmaker's guide to working with indigenous people, culture and concepts*. Sydney: Screen Australia. www.screenaustralia.gov.au/getmedia/e601f1b9-0394-4c83-9a62-c20939d9ab30/Indig_Protocols.pdf. Accessed December 29, 2013.

Jasanoff, S. 2010. A new climate for society. *Theory, Culture & Society* 27: 233–253.

Johansson, T. 2002. Lighting the campfire: The creation of a community of interest around a media company. *International Journal on Media Management* 4: 4–12.

Johnson, M. 1992. *Lore: Capturing traditional environmental knowledge*. Darby, PA: DIANE Publishing.

Kahan, D. 2010. Fixing the communications failure. *Nature* 463(7279): 296–297.

——, H. Jenkins-Smith, and D. Braman. 2011. Cultural cognition of scientific consensus. *Journal of Risk Research* 14: 147–174.

Karimova, S. 1980. *The increasing of labor activity of the women in the development of agriculture in Tadzhikistan* (in Russian). Dushanbe, Tajikistan: DONISH publishing company.

Kassam, K.-A. 2009. Viewing change through the prism of indigenous human ecology: Findings from the Afghan and Tajik Pamirs. *Human Ecology* 37(6): 677–690.

Kayumov, A. and I. Rajabov. 2010. Glaciers – Water resources of Tajikistan in condition of the climate change. Dushanbe, Tajikistan: State Agency for Hydrometeorology, Government of the Republic of Tajikistan. http://aoa.ew.eea.europa.eu/virtual-library-viewer/answer_2335310034. Accessed December 28, 2013.

Kindon, S. 2003. Participatory video in geographic research: A feminist practice of looking? *Area* 35: 142–153.

MacLean, K. and E. Woodward. 2013. Photovoice evaluated: An appropriate visual methodology for aboriginal water resource research. *Geographical Research* 51: 94–105.

Makhmadaliev, B., A. Kayumov, V. Novikov, N. Mustaeva, and I. Rajabov. 2008. The second national communication of the Republic of Tajikistan under the United Nations Framework Convention on Climate Change. Dushanbe, Tajikistan: State Agency for Hydrometeorology, Government of the Republic of Tajikistan.

Maynard, N.G., A. Oskal, J.M. Turi, S.D Mathiesen, I.M.G. Eira, B. Yurchak, V. Etylin, and J. Gebelein. 2011. Impacts of Arctic climate and land use changes on reindeer pastoralism: Indigenous knowledge and remote sensing. In *Eurasian Arctic land cover and land use in a changing climate*, eds G. Gutman and A. Reissell, 177–205. Dordrecht, Netherlands: Springer.

Michaels, E. 1986. *The aboriginal invention of television in Central Australia, 1982–1986: Report of the Fellowship to Assess the Impact of Television in Remote Aboriginal Communities*. Canberra, Australia: Australian Institute of Aboriginal Studies.

Mustonen, T. and K. Mustonen. 2010. *At the gates of the sun: A Snowchange book of images from indigenous nomadic communities of the Republic of Sakha-Yakutia, Russia*. Tampere, Finland: Snowchange Cooperative.

Nabhan, G.P., K. Wilson, O. Aknazarov, K.-A. Kassam, L. Monti, D. Cavagnaro, S. Kelly, T. Johnson, and F. Sekacucu. 2012. Agrobiodiversity shifts on three continents since Vavilov and Harlan: Assessing causes, processes, and implications for food security. In *Biodiversity in agriculture: Domestication, evolution, and sustainability*, eds P. Gepts, T.R. Famula, R.L. Bettinger, S.B. Brush, A.B. Damania, P.E. McGuire, and C.O. Qualset, 407–425. Cambridge: Cambridge University Press.

Nabieva, R. 1999. *The role of women in society* (in Russian). Dushanbe, Tajikistan: IRFON publishing company.

Nakashima, D.J., K. Galloway McLean, H.D. Thulstrup, A. Ramos Castillo, and J.T. Rubis. 2012. *Weathering uncertainty: Traditional knowledge for climate change assessment and adaptation.* Paris: UNESCO, and Darwin: United Nations University. www.ipmpcc.org/weathering-uncertainty-traditional-knowledge-for-climate-change-assessment-and-adaptation/. Accessed December 29, 2013.

Nerlich, B., N. Koteyko, and B. Brown, B. 2009. Theory and language of climate change communication. *Wiley Interdisciplinary Reviews: Climate Change* 1: 97–110.

Nightingale, A.J. 2003. A feminist in the forest: Situated knowledges and mixing methods in natural resource management. *ACME: An International E-Journal for Critical Geographers* 2: 77–90.

NordNordWest. 2008. File:Tajikistan location map.svg – Wikipedia, the free encyclopedia. *Wikipedia.* http://en.wikipedia.org/wiki/File:Tajikistan_location_map.svg#file. Accessed March 2, 2014.

PALM Project. 2011. Sustainable Land management in the High Pamir and Pamir Alai Mountains (PALM). www.ehs.unu.edu/palm/article/read/western-pamirs. Accessed December 29, 2013.

Parry, M.L., O.F. Canziani, J.P. Palutikof, P.J. van der Linden, and C.E. Hanson, eds. 2007. *Climate change 2007: Impacts, adaptation and vulnerability*. Cambridge: Cambridge University Press.

Petherick, A. 2013. Oblivious money-men. *Nature Climate Change* 3: 96–97.

Pink, S. 2007. *Doing visual ethnography*. London: Sage.

——, ed. 2012. *Advances in visual methodology*. London: Sage.

Radcliffe, S.A., E.E. Watson, I. Simmons, F. Fernández-Armesto, and A. Sluyter. 2010. Environmentalist thinking and/in geography. *Progress in Human Geography* 34: 98–116.

Ravetz, J.R. and Z. Sardar. 1996. *Cyberfutures*. London: Pluto Press.

Rocheleau, D. 1995. Maps, numbers, text, and context: Mixing methods in feminist political ecology. *The Professional Geographer* 47: 458–466.

Root, M. 1993. *Philosophy of social science: The methods, ideals, and politics of social inquiry*. Oxford: Blackwell.

Russell-Smith, J., D. Lucas, M. Gapindi, B. Gunbunuka, N. Kapirigi, G. Namingum, K. Lucas, P. Giuliani, and G. Chaloupka. 1997. Aboriginal resource utilization and fire management practice in western Arnhem Land, monsoonal northern Australia: Notes for prehistory, lessons for the future. *Human Ecology* 25: 159–195.

Sardar, Z. 1995. Alt. civilizations. Faq cyberspace as the darker side of the West. *Futures* 27: 777–794.

Soukup, K.T. 2006. Travelling through layers: Inuit artists appropriate new technologies. *Canadian Journal of Communication* 31(1): 239–246. www.cjc-online.ca/index.php/journal/article/view/1769/1889. Accessed January 12, 2014.

Srinivasan, R. 2007. Ethnomethodological architectures: Information systems driven by cultural and community visions. *Journal of the American Society for Information Science and Technology* 58: 723–733.

——. 2012. Re-thinking the cultural codes of new media: The question concerning ontology. *New Media & Society* 15: 203–223.

Stam, R. and E. Shohat. 1994. *Unthinking Eurocentrism: Multiculturalism and the media.* London: Routledge.

Terry, G. 2009. No climate justice without gender justice: An overview of the issues. *Gender & Development* 17: 5–18.

Tricia Fragnito, S. and J. Lewis. 2005. Aboriginal territories in cyberspace. *Cultural Survival Quarterly* 29 (2, Summer). www.culturalsurvival.org/publications/cultural-survival-quarterly/canada/aboriginal-territories-cyberspace. Accessed December 29, 2013.

Turner, T. and N. Lantz. 2003. Traditional phenological knowledge (TPK) of aboriginal peoples in British Columbia. *Journal of Ethnobotany* 23: 263–286.

UNEP (United Nations Environment Programme). 1992. 46th Sess.; Agenda Item 21; U.N. GAOR; UN Doc A/Conf.151/26.

UNPFII (United Nations Permanent Forum on Indigenous Issues). 2009. State of the world's indigenous peoples. United Nations Secretariat Permanent Forum on Indigenous Issues, Department of Economic and Social Affairs. New York: United Nations. www.un.org/esa/socdev/unpfii/documents/SOWIP_web.pdf. Accessed December 28, 2013.

Wilson, K.M. 1995. Mass media as sources of global warming knowledge. *Mass Communication Review* 22: 75–89.

Wilson, P. and M. Stewart. 2008. *Global indigenous media: Cultures, poetics, and politics.* Durham, NC: Duke University Press.

Wilson, S. 2008. *Research is ceremony: Indigenous research methods.* Winnipeg, MB: Fernwood Publishing.

Yibarbuk, D., P.J. Whitehead, J. Russell-Smith, D. Jackson, C. Godjuwa, A. Fisher, P. Cooke, D. Choquenot, and D.M.J.S. Bowman. 2001. Fire ecology and Aboriginal land management in central Arnhem Land, Northern Australia: A tradition of ecosystem management. *Journal of Biogeography* 28: 325–343.

Zakrieva, M. 2001. *Histography of female issue in Tadzhikistan* (in Russian). Dushanbe, Tajikistan: Moscow State University publishing office.

Zurba, M. and F. Berkes. 2013. Caring for country through participatory art: Creating a boundary object for communicating Indigenous knowledge and values. *Local Environment: The International Journal of Justice and Sustainability*, published online, 16 pp. http://dx.doi.org/10.1080/13549839.2013.792051. Accessed December 29, 2013.

12 Conclusions

Advancing multi-disciplinary scholarship on gender, water, and environmental change through feminist political ecology

Stephanie Buechler, Anne-Marie Hanson, Diana Liverman, and Miriam Gay-Antaki

As Isabel Dyck has argued, "We need [to pay] close attention to the spaces of everyday life to keep women visible in rapidly changing world conditions, where their activities tend to slip into the shadows of dominant models in the literature" (Dyck 2005: 234). One of the ways to ensure that women's strategies are made visible in the face of these dynamic world conditions is by enhancing opportunities for women's voices to reach a larger public. The authors who contributed to this book focus their applied research on the everyday spaces that women inhabit and actively shape, especially as these spaces are impacted by climate change and degradation of water resources. Their work stresses the intersections of gender, ethnicity, race, social class, and other social categories as they relate to environmental management and political power dynamics. These same social differences impact ecologies and the ways women and other marginalized groups participate in resistance to hegemonic decision-making in order to proactively influence livelihoods, well-being, and rights around water resource issues. Within their chapters, authors include some of the exact words of the women and men involved in various research projects and development programs in order to better convey their strategies and goals.

The authors engage the diverse disciplines of Geography, Sociology, Anthropology, Public Policy, Women's Studies, Economics, Environmental Studies, Development Studies, Latin American Studies, Landscape Architecture, Media Studies, and Environmental Philosophy. Each author works in a multi-disciplinary manner to analyze ecological and political issues related to women and water. Within their various disciplines, all authors push the boundaries of theory and empirical research. At the same time, feminist political ecology, intersectionality, and environmental justice framings unify their contributions to this book. It is through this multi-disciplinary approach to women and water issues that the spaces of everyday life are uncovered. This multi-disciplinarity also helps to reveal—from different scales and modes of understanding—how the environmental context of global change both shapes and is shaped by individuals, by particular social groups, by entire communities, and by cross-regional collaborations.

This concluding chapter is organized around the main findings of the book, which include the following converging themes: 1) linking water and global environmental change; 2) women's proactive roles; 3) the importance of place and scale; 4) women, environment, and economy; 5) methods and theories for the study of gendered intersections with ethnicity, race, and social class; 6) multi-scalar policy and programmatic action; and 7) visions for the future of applied research and action on women, water, and global environmental change.

Linking water and global environmental change

Overall, this volume demonstrates the precariousness of water resources and the consequences of this precariousness in terms of water availability, accessibility, and quality for human communities globally. Water is a fundamental necessity for all life on Earth. Yet the precious 1 percent of freshwater available on Earth for human, animal, and plant life is becoming increasingly degraded and scarce. As Lankao and Smith point out in the most recent IPCC (Intergovernmental Panel on Climate Change) report: "[c]limate stresses occur alongside other anthropogenic influences on ecosystems, including land-use changes, non-native species, and pollution, and in many cases will exacerbate these pressures" (Lankao and Smith 2014: 4). An example of anthropogenic effects on freshwater and livelihoods is the case of copper mining in northwestern Mexico that caused the massive sulfuric acid contamination of the Sonora River in August 2014. Although largely unpublicized in the international media, over 40,000 m^3 of toxic wastewater from copper mines was leached into the river, leaving people without water in communities along the river in seven municipalities and in Hermosillo, the capital city of Sonora. Fish, livestock, and other animals died, and the federal government was forced to prohibit the use of wells used for drinking water in the capital city of 857,225 people and in rural and peri-urban areas. This also affected water resources used for farming and agricultural livelihood activities, such as women's cheese production, all along the Sonora River (López Ruelas 2014; Salud Sonora 2014).

Rapidly changing climate and demographic factors also impact freshwater sources. Freshwater quality and quantity in rivers, streams, ponds, wetlands, lakes, and other surface water bodies are reduced by the increasing variability of climatic factors such as precipitation, temperatures, and wind patterns. These water bodies are also strained by the ever-growing human appropriation of freshwater for urban and rural drinking water, agricultural production, industrial and institutional purposes, as well as other indoor and outdoor residential uses. Freshwater availability and quality in springs and wells are also affected by climatic variability as groundwater levels are directly linked to surface water levels through the percolation of surface water into the ground (Arnell 2004; Combes 2003). In the Arctic, the effect of higher temperatures is stronger than anywhere else in the northern hemisphere or the globe. As warmer conditions melt surface snow and ice, the land and water underneath absorb more of the sun's radiation, leading to even more snow and ice melt in a process termed arctic amplification (Harriss 2014). This process, in turn, is also linked to rapid changes in Arctic peoples' ways of life, including their transportation methods over the ice and the distribution of species in their hunting and fishing grounds (Herman-Mercer et al. 2011).

As the 1 percent of available freshwater is increasingly under pressure from human use, contamination, and climate change, the rest of the water on Earth is found in the oceans—which are also severely impacted by human activities. More than 90 percent of the excess CO^2 in the atmosphere is soaked up by the oceans. When CO^2 is dissolved in water, it causes ocean acidification. A report released in 2013 from the University of Western Australia's Oceans Institute, found that global warming could cause oceans to become 170 percent more acidic by the end of the century, the fastest rate of acidification in the past 300 million years (Milman 2014). Acidification effects, in addition to rising sea levels, will greatly impact urban and rural populations living near sea coasts. This includes livelihoods dependent on fish and other marine resources. Rising sea levels are threatening communities with homes and workplaces near coastal infrastructure like roads and bridges. In some areas, agricultural lands are becoming submerged. These rising sea levels also threaten groundwater tables further inland due to saline intrusion (Webb and Howard 2011).

Proactive roles of women

The authors in this book stress that women are often very active in their communities' efforts to mitigate the negative effects of environmental change on water resources. Women are also increasingly proactive in collaborative work with other groups beyond the confines of their own communities to address human-caused contamination and other human-engineered water problems. While all chapters address the often disadvantaged position from which women take on such proactive roles, a major goal of this book is to move from a focus on absences or the silencing of gendered and subaltern knowledges to stress the politics of emerging social movements and collaborative action (Escobar 2004; Gibson-Graham 2006; Gibson-Graham and Roelvink 2009). The innovative examples provided within the case studies presented here demonstrate women's and indigenous groups' engagement in movements for political, environmental, and social change, and are useful models for other areas facing similar problems. They contribute to the literature on gender and development (Coles et al. 2015; Momsen 2010; Visvanathan et al. 2011) and women, the environment, and development (including sustainable development) (Braidotti et al. 1994; Cruz-Torres and McElwee 2012; Dankelman 2002; Osnes 2014).

The types of social action that are documented in this book include a broad array of actors in diverse contexts, all with the common goal of improving water-related obstacles to their livelihoods, health, and well-being. Lahiri-Dutt documents the case of women who, after negotiations fail, go with their fellow villagers to the mine; this mining company is diverting the spring water the women had always used for drinking water and household use to the coal mine's operations and to residential areas for managerial staff. Women courageously stand in front of the machines to halt mining operations. In Kambic's chapter on Los Angeles, women offer more democratic alternatives to large-scale, de-personalized urban water management through the creation of small-scale, locally appropriate organizations that use rainwater and greywater. The city is also creating public parks that help to connect low-income communities to the Los Angeles River and to native vegetation.

Women are also politically active in water governance issues, despite many social and cultural barriers to women's equal participation in formal politics. In northeast Brazil, women are active in watershed management at all levels, although poor women of color experience barriers to participating in regional level management according to Moraes. Buechler's case illustrates how women in northern Mexico are pressuring government officials for greater recognition of and support for the agricultural processing (cheese production) activities they engage in and for the non-irrigated areas where most of the dairy cows are pastured. Perkins and Walker show that women and girls are highly engaged in learning about watershed issues, and that they translate the knowledge they gain from international collaborations into work in their urban communities. In the resettlement areas along the Nile River in Egypt, Najjar highlights the important ways that women are participating in local water management, but also discusses how women still lack the respect of government, community water managers, and agricultural development professionals.

Women's activism plays an important role in community water issues including the protection of local health, water resources, and water traditions from local to global scales. Hanson's research in coastal Yucatán, Mexico sheds light on a community initiative led by women who address problems related to water pollution by garbage and rising sea levels, and shows how their work is replicated throughout coastal Yucatán and beyond. Canadian Tlingit women's prominent roles as activists in the social movement *Idle No More* work to defend indigenous traditions and water resources in the Yukon lakes and are portrayed in Hayman's collaborative case study. The Pamiri women in Williams and Golovnev's study speak out in a film on their adaptation strategies to climate change and melting glaciers in order to allow international policymakers, Pamiri peoples in other communities, academics, and others to listen and to learn from their experiences.

Place and scale

The studies in this book all, either explicitly or implicitly, speak to the importance of place or location in relation to water access, water scarcity, and water contamination. Social and physical locations are key factors that shape the multi-level connections between global and local scales and between urban and rural spaces. These locations, in turn, affect decision-making strategies and hierarchies of power with relation to gender and water resources.

Urban political ecologists in particular have addressed the problematic tendency to categorize environmental politics, subjectivities, and material practices by location. As Keil notes, "the material and symbolic, the natural and the cultural, the pristine and the urban are not dual and separate realities but rather intertwined and inseparable aspects of the world we inhabit" (2003: 728). Despite this scholarly recognition that the conceptual division between urban and rural spaces is inherently fragile (Ávila Sanchez 2005; Buechler 2002; Harvey 1996; Hodson and Marvin 2010; Jiménez-Solares 1998; Kearney 1996; Williams 1973), political ecology research often continues to be divided under the separate rubrics of 'urban' or 'rural' analyses (Braun 2005; Brockington et al. 2008; Heynen et al. 2005; McDonald et al. 2009).

This urban–rural divide is increasingly problematic when addressing the connections of gendered social networks and globalization processes that both control and unevenly distribute water to marginalized groups. Urban–rural connections are especially visible in terms of water quality, availability, and accessibility, and in the consequences of water resource deterioration.

Methodologically and theoretically, this urban–rural divide needs to be bridged. The empirical, methodological, and theoretical contributions of the chapters to this topic are multi-faceted. Kambic's chapter on the city of Los Angeles reminds us that while some dynamics are place-based, others are more ubiquitous. This is evidenced by the growing inaccessibility of urban populations to public water points such as drinking fountains and the limited access to recreational areas such as parks with water in the landscape. These dynamics illustrate a construction of urban space that does not respect nature, but rather subverts it through the re-channeling of rivers and the separation of vegetation from the river. Grassroots groups address these problems by re-linking the river water flowing within the city to the watershed. This, in turn, allows native vegetation to grow in the river channels and has led to the creation of river parks all along the river. This represents the linking of rural, peri-urban, and urban places throughout the watershed. Kambic starts from the vantage point of landscape architecture, then includes geography and sociology to help connect the dots between the watershed, the city, and grassroots groups. Her use of feminist political ecology and intersectionality frames address the connections between the micro-levels of the household, to the city-scale, to the macro scale of the region and watershed.

Williams and Golovnev incorporate both rural and urban locales in their visual methods to show how, in the mountains of the Pamir region of Tajikistan, rivers and glaciers are affected by climate change and are altering women's lives across the life cycle from the young to the old. They film and conduct interviews in rural and urban spaces across the watershed, enabling a glimpse into differences and common experiences with climate change that could act to bind the peoples in this region together. They show the film in many different parts of the Pamir region, in the full spectrum between rural and urban, to help spur local discussion and action and to encourage better climate change communications among peoples across the region.

Paradoxically, in semi-arid regions of northern Mexico (Buechler) and Egypt (Najjar), as well as in lake- and river-dotted Yukon, Canada (Hayman) and the river and spring-abundant Lesotho highlands (Braun), climate-induced changes in rainfall patterns and growing urban populations are leading to diversions of available freshwater into dams (often with the effect of drying up springs and wells). These diversions of water are often for distant urban use (including water supply and energy production). These authors use feminist political ecology to assess the gender, ethnicity, and class dimensions surrounding questions of whose ways of life are most affected by depleting and increasingly polluted water sources. Careful, in-depth ethnographic studies were conducted over time to illuminate processes of change. These ethnographies were combined with secondary research on the broader ecological, cultural, and political contexts to show the interconnections of water and people across urban and rural landscapes.

Hanson's and Lahiri-Dutt's contributions show clearly the impacts of urban consumer demand from regional and global markets on marginalized populations. Hanson's study of towns in coastal Yucatán uses ecological oral histories to reveal the negative ramifications of international consumption patterns that produce plastic waste that, with intensified storms and rising sea levels, is increasingly washing ashore and littering residential and tourist areas. Hanson argues that these towns do not receive the level of waste services that large cities in Mexico do (although large cities also have major garbage issues). This lack of adequate waste management in coastal areas is just one effect of the government's and outsiders' images of these towns as rural when they actually now have more urban than rural characteristics. Lahiri-Dutt's case study of a coal mining village in the eastern state of Jharkhand, India, illustrates how an industry geared toward meeting the global and regional demand for energy in urban areas has deep effects on natural resources in rural places. Rights-based and feminist political ecology approaches are used to illustrate the effects of a coal mining company's operations on villagers' rights to water, land, and other resources, and the uneven nature of these effects across social groups in coal mining villages.

Perkins and Walker deftly illustrate that in climate mitigation and adaptation related applied research, rural and urban places can be connected in the minds and actions of people of all ages (with a focus on girls and women). The projects primarily include trainings related to watershed management and include communities in disparate locations *vis-à-vis* cities in the Global South and in the North. The watershed scale of the projects helps to show the connections between dispersed communities and also aids in unveiling the similarities that exist between regions as different as Brazil and Canada and between Mozambique, South Africa, and Kenya. The utility of feminist political ecology for applied work that includes the watershed scale is amply illustrated by these authors.

Connecting women, environment, and economy

The physical and social locations of water users are also linked through their economies. Feminist political ecology "joins theory, politics and practice by working from case studies, with a focus on complex interlinkages among gender, environment, culture, and economy 'in place'" (GESEC 2011: 43). The incorporation of water into market-based relations creates deeper inequalities by gender, race, ethnicity, and class and creates new social divisions (Ahlers and Zwarteveen 2009; Sankey and Cárdenas 2013; Truelove 2011).

This book builds on these insights by feminist political ecologists and other feminist researchers to help expose the ways in which climate change and other types of global environmental change impact water resource-related market dynamics. Chapters address these issues to show the intricate ways in which climatic change alters economic relations.

In Buechler's chapter on semi-arid northern Mexico, climate change creates barriers to the access of resources such as water, but also to the access of crops that contain 'virtual water.' Virtual water is the water used to produce agricultural products, which in Buechler's case study includes alfalfa purchased from another region (where it is grown by large producers with, at least at present, sufficient finances to deepen their wells) and used to feed cows to obtain the milk used for cheese production. In Hayman's chapter on Canada, transformations in economic relations are evinced through climate-induced change in rainfall patterns that lead to the diversion of rivers and lakes for often distant urban use (including water supply and energy production for urban populations) and agriculture. Braun's study also examines economic changes wrought by the construction of a dam in Lesotho. In her case, the large-scale dam was constructed without considering the use value of the river for the local people; the only value considered was the economic valuation of the dam's utility as an economic instrument to boost national GDP.

Hanson examines the intersection of globalized markets (related to consumption patterns) with climate change-induced sea level rise and more severe tropical storms, and notes the deleterious effects on livelihoods in coastal communities (and their innovative responses). Lahiri-Dutt's chapter shows that mining (in this case, climate-harmful coal mining) leads to new gender cultures where cash becomes a status symbol leading to power; since cash is mainly accessed by men, women become further marginalized socially, economically, and politically, and the water sources they depend on become increasingly degraded. Najjar documents Egyptian women's successful use of globalized tastes in urban markets to market their cactus fruit that they now produce in the areas they were relocated to due to sea level rise. Perkins and Walker's watershed protection projects with women and girls were based on the premise that it is less costly for communities to prevent flooding and other effects of climate change than to have to rebuild after climate-related events.

Intersections of gender with ethnicity, race, and social class in environmental contexts

Theoretical implications of the case studies

Feminist political ecology understands issues of power by stressing how everyday practices from local to global sites maintain unequal power relations. These practices might include the silencing of non-English speaking voices from climate negotiations and climate assessments or the token acknowledgment of gender in climate discourses without addressing material gender inequities. Contesting these everyday inequities requires counter narratives, solidarity, and struggles to destabilize the structures and discourses that maintain control over the resources, political decisions, and peoples that create vulnerability, foster high carbon emissions, or overlook power in climate policy (Bee 2013; Bee et al. 2013).

This book points to the continued and invaluable importance of everyday practices and how such practices are linked to scale in gendered environmental analysis. All chapters draw from feminist political ecology frameworks and they stress the integration of multiple scales of analysis from the intimate to the international. The chapters by Hanson, Moraes, Williams and Golovnev, Kambic, Najjar, and Hayman illustrate the intricate links between the micro scale of the individual woman and man, household and community, to the macro-level of the city, region, or watershed. Their scalar analysis sheds light on women's and men's different involvement in political processes related to water and environmental change, and the significance of this differential involvement for addressing the particular challenges existent at each scale. This scalar analysis allows the authors to highlight the importance of scaling up to achieve broader policy impacts and the need for more coordination between different levels of policies and programs, from grassroots to global levels.

Several chapters also explore intersectionality and environmental justice frameworks through diverse disciplinary understandings of political ecology and feminist political ecological analysis. From a landscape architecture standpoint, Kambic shows how intersectionality frames can reveal the effects of top-down urban design on urban water infrastructure. She focuses on the water access points that exclude the poor and minority groups, and points to visions of a feminist-framed future that includes open access to swimming pools, drinking fountains, and parks, and the utilization of local resources that are managed at the micro-level such as urban produced greywater and rainwater. From intersectionality and environmental justice perspectives, Braun's study shows that the rural and urban poor are disproportionately affected by the construction of dams that flood agricultural land and whole communities. Poor women are the most affected due to their roles as water provisioners, wild plant harvesters, and collectors of soils, wood, and other resources and due to their limited ability to enter paid labor which was mainly given to foreign workers. Braun's and Lahiri-Dutt's studies also use a rights-based approach related to water and land resources which strengthens their feminist political ecology and intersectional analyses of gender, rural status, and class-based social inequities.

This volume also highlights issues of race and ethnicity in intersectionality analyses. The concept of intersectionality reminds us that a narrow focus on gender may constrain our understanding of women's and men's experiences with environmental change. Many forms of social differences and oppression interact with gender, such as class, race, ethnicity, and sexuality, and are simultaneously important in determining responsibility, vulnerability, or governance of water and climate change issues (Kaijser and Kronsell 2014; Mollett and Faria 2013; Nightingale 2011). Identities are rarely fixed but are fluid and contingent within a changing environment. Moraes examines the influence of race, class, and educational levels as factors affecting women's inclusion in watershed management to find that gender, race, and class intersect to act as barriers to women's participation in higher levels of watershed management in arid Northeast Brazil. These dynamics will become even more problematic with the predicted future dryness that will require more effective rainwater catchment and inclusion of local knowledge to manage the very little rain that is expected to fall. Hayman argues that gender needs to be de-emphasized in some areas of research within feminist political ecology. She illustrates how gender is understood very differently within some indigenous cosmologies than in western cultures. Among the Tlingit of northern Canada, while gender does shape experiences with ecological change to some degree, more important is the relationship between individuals and water bodies as expressed through language and storytelling. Place names and narrative socio-environmental stories are central to Tlingit people's identity and feelings of continuity in the context of environmental change.

Perkins and Walker emphasize local, regional, and global collaborations for social-environmental change and show how grassroots feminist political ecology can help to shed light on these collaborations. Multi-sited examples offered by Perkins and Walker also enable consideration of the types of projects that might support and enable the next generation of leaders in micro-level and watershed-level governance. Their study pays special attention to the need to incorporate girls and women, especially those living in communities economically and geographically vulnerable to climate change. Their use of feminist political ecology is innovative in showing how knowledge flows can occur across several scales. Their study also points to the necessity of including age in feminist political ecology and in intersectionality analyses. Theory could then inform empirical work and ultimately policy regarding age-related challenges in water resource and global change issues. Youth, children, and the elderly have been largely invisible in water resource-environmental change accounts and analyses. Too few programs and policies have been tailored to these social groups, particularly elderly women and girls. By including age as a focus of analysis in intersectional and feminist political ecology studies, issues such as the passing down of gendered ecological knowledge from one generation to another, the development of new ways of learning, and adaptation strategies to emerging environmental threats and opportunities can be better incorporated into theory, empirical research, and practice.

Buechler's study of Sonora, Mexico, and Najjar's of Egypt both point to the need for feminist political ecology theory to incorporate a stronger focus on local ecologies. Gendered ecological knowledge is also better understood when emphasis is placed on local and regional ecologies. Buechler argues for the combining of socio-ecological systems (SES) analysis, feminist political ecology, and political ecology in order to address this lack of adequate attention to dynamic ecological processes while still incorporating a focus on gendered power and decision-making structures.

Methodological advances

Most authors in this book used ethnographic methods for their studies and came to a similar conclusion: the importance of hearing different voices related to global environmental change experiences. Bringing the voices of women to a larger public is a major emphasis in all chapters even though the authors come from very different disciplines. Piers Blaikie recommended in his lecture 'Should some political ecology be useful?': "we need to include the politics of this science . . . alternative voices which express alternative knowledges of the environment should be sought, heard and recorded" (2012: 237). Townsend and her collaborators argued in *Women's Voices from the Rainforest*:

> As feminists, we want to represent the lives of . . . women and to give them some opportunity to represent themselves, their problems and their solutions, while recognizing that only partial success in these aims is possible . . . There will be no solution without these voices.
>
> (Townsend 2005: 14)

Feminist researchers across disciplines underscore that for the most part knowledge has been co-opted by a euro/male-centric view and is inherently hierarchical and that research practices run the risk of reifying these hierarchies. Feminist methodological debates have engaged with issues of power in the research process (Elwood and Martin 2000; Gibson-Graham 1994), as well as power relations embedded in participatory development including water resource-related development projects (Agarwal 2001; Aladuwaka and Momsen 2010; Cleaver and Hamada 2010; Walker et al. 2007). Feminists have sought to destabilize knowledge hierarchies by collaborating with rather than 'studying' those 'researched' (Nagar 2006). Using a feminist approach to employ standard research methods can work to denaturalize gendered differences.

Mixed-methods can elucidate different ways of knowing which in turn can reveal important aspects of lived experience (Nightingale 2003; Rocheleau 1995). Combinations of ethnography (Pillow and Mayo 2011), gendered mapping (Alagan and Aladuwaka 2014; Katz 2001), co-writing (Nagar 2006), surveys (Stewart and Cole 2007), photo-voice (Giritli-Nygren and Schmauch 2012), Skype calls, videos (Khamis et al. 2009; Zaman and Reichert 2014), and other audio-visual collaborations lead to more inclusive projects in terms of the types of questions asked and also create propitious spaces for bi-directional flows between researcher and research project participant. These types of methods pay particular attention to how knowledges are produced and can help empower women as agents in responses to global environmental change. Yet, as practitioners and researchers of development note, while a focus on research *process* is important, the outcomes or *products* of research may be equally or more important for participants. One of the goals of mixed-methods and participatory research methods is that they will *produce more appropriate data* for residents and government agencies than 'colonial' or simple data-extracting research methods (Fortmann 2008; Shanley 2006).

Climate research and policy has mostly consisted of unilateral transfers of knowledge from North to Global South. Climate change is framed in such a way that it deters many actors across the globe from actively participating in decision-making. Through feminist approaches, knowledge building becomes a relational process, opening spaces for different types of knowledge to be legitimized and included within the techno-scientific framework where climate change has lately resided.

Most academic and policy analyses of the environment, however, do not come to these conclusions concerning the importance of participatory methods or the explicit inclusion of women and other marginalized voices. This oversight has important ramifications for the development of better policies and programs. Supporting women's and indigenous people's equal voice in decision-making linked to water and environmental change is needed in the home, in the community, and at the national and global levels. However, in order to ensure that greater equity is achieved regarding this decision-making, stronger focus on the types of methods used to elicit these groups' opinions and experiences is necessary.

In this volume, several innovative methodological tools are utilized; most involve the use of the exact words spoken by the women in the communities involved in the research project. Storytelling, ecological histories, narratives, oral testimonies, community stakeholder meetings, visual methods such as collaborative video, in-depth interviews and follow-up interviews, and participant observation are all methods used by the authors to help bring the voices of the women to the larger public through the written word. By providing space for women's views to be heard, we hear their accounts of the past and the changes that have occurred, their accounts of current initiatives they are involved in, and visions they have for a different future.

Collaboration is offered by feminist political ecology as a way to work with (rather than on or about) others to shape research that is relevant to, for example, local women activists, to undertake research together and communicate the results (Nagar and Ali 2003; Sharp 2005; Sundberg 2004). Collaborative approaches have considerable potential in studies of community-based adaptation, vulnerability mapping, and understanding gender in international climate governance. In this volume, nearly all of the studies were collaborative in nature. In particular, Perkins and Walker describe the participatory methodologies used in sister watershed projects across international boundaries. Through direct training and participatory research processes that elicited the voices of the young and the old, women were able to share the challenges and the benefits of their participation in watershed development programs with other women involved in similar programs in faraway places and across cultures.

Hayman's study with the Tlingit peoples of the Yukon, Canada and Williams and Golovnev's study with the Pamiri peoples of Tajikistan also show the value of methods such as collaborative writing and filmmaking. For these studies, community members were consulted with respect to the questions asked, the analysis of climate issues, and in the actual write-up of the research. Hayman accomplished this through the inclusion of a Tlingit woman and man as community consultants and through the use of in-depth, in-person interviews, Skype calls, and email discussions from her home near Munich, Germany to the community consultants' homes in the Southern Yukon Lakes in the Yukon Territory, Canada. In the case of the research with urban and rural women in the Pamir Valley, a collaborative video was produced; this video is owned and distributed by those who appear in the video as well as by the bi-national scholars and their respective institutions.

Policy and programmatic implications of the studies

Peter Walker's argument regarding political ecology still holds today, that is, that there is need in political ecology:

> for better articulation and coordination as an intellectual community to put the vast wealth of knowledge generated within the subfield to work for tangible problem-solving . . . a failure to fully and energetically engage policy at a time when society and the planet urgently need their perspectives is a political act with profound implications.
>
> (Walker 2006: 392–393)

All of the case studies point in some way to the need for policy that better recognizes and rewards the positive contributions of women to address the negative ramifications of global environmental change. In particular, water-climate linkages analyzed in the chapters include decreasing rainfall resulting in declining water levels in rivers and groundwater tables, desertification, carbon emissions, water pollution from mining and garbage, rising sea levels, melting ice sheets, and increasing water and air temperatures and resultant evaporation. Research has long shown differences between men and women in their knowledge of and responses to climate change. Surveys in the U.S. find that women are more knowledgeable about climate change, more confident and concerned that climate change is happening and harming people, and more likely to take actions to reduce energy use (Goldsmith et al. 2013; McCright 2010). Weichselgartner and Kasperson found that women decision-makers and practitioners were more likely than their male colleagues to accept studies on global environmental change that relied more on qualitative than on quantitative research or studies that incorporated theory (2010: 271).

The IPCC reports released in 2014 raise questions about the gender dimensions of climate impacts and vulnerability, but generally overlook the positions or views of women as they relate to emissions reductions or climate policies. Attention to gender has increased modestly within climate governance, with a few more women now represented at climate negotiations, with women and gender issues advocated by non-profit, governmental, and grassroots organizations, and as more applied researchers are working on gender and climate change governance (Petherick 2013; Seager 2009; Tuana and Cuomo 2014; Wamukonya and Skutsch 2002). For example, in terms of capacity building, The Troika+ of Women Leaders on Gender and Climate Change is a group of high-level women leaders funded by the Mary Robinson Foundation-Climate Justice (MRFCJ) to meet and work together to strengthen women's leadership skills to draw attention to the gender dimensions of climate change (www.mrfcj.org). With Rockefeller Foundation funding, the United Nations climate change secretariat or United Nations Framework Convention on Climate Change (UNFCCC) launched 'Momentum for Change: Women for Results' which grants funding for 'lighthouse activities,' which are projects designed to increase women's resiliency.

The interests of women at climate negotiations are represented by the participation of a number of international non-governmental organizations—such as the Global Gender and Climate Alliance—and by several events focusing on gender in recent years. The UNFCCC web page now has a section on gender stating:

> It is increasingly evident that women are at the centre of the climate change challenge. Women are disproportionately affected by climate change impacts, such as droughts, floods and other extreme weather events, but they also have a critical role in combatting climate change.
>
> (UNFCCC 2014)

UNFCCC chief, Christiana Figueres, has urged that:

> An important part of leadership is to inspire people to go beyond where they think they can – and motivate them to think outside the box. A critical nexus is energy, water and food . . . and when you look in the middle of the triangle – what's there? Women! My vision is that we should have 50-50 participation in all decision making, 50% women and 50% men.
>
> (Quoted in King 2012)

The chapters in this volume all include a look ahead into the future, and this future includes more uncertainty regarding climatic conditions. The Fifth Assessment Report of the IPCC (Lankao and Smith 2014) is even more confident than previous reports that the planet is warming, that the causes of the warming are anthropogenic, and that the impacts will vary with the vulnerability of different places and people. The causes of anthropogenic climate disruptions are numerous, and all of these disruptions have consequences for uneven distribution and access to water resources.

Highly industrialized countries account for an average five times greater per capita emissions than countries in the Global South. Energy production emits about one-third of all carbon. The burning of coal, the extraction of which is addressed in Lahiri-Dutt's chapter, has the largest emissions. In the next 20 years, several hundred million people will receive electricity produced by coal unless current trends are altered. India has the world's fifth largest coal reserves in the world, and is currently the third largest producer; of the 50 major mining districts of India, almost half are in predominantly tribal (native peoples) districts (Nijhuis 2014). Increased use of coal to fulfill global energy demands is also leading to uneven distribution of pollution on freshwater resources that are already under pressure from other human uses. Methane is another very harmful greenhouse gas. Methane gas emissions from rice and livestock production as well as food waste are also now recognized for their considerable contribution to greenhouse gas emissions (Johnson et al. 2007). On a shorter timescale of a few decades, methane is more powerful than carbon dioxide in terms of its heat-trapping potential and over the course of 100 years it is equivalent to carbon dioxide as a greenhouse gas (Howarth et al. 2011).

Program development and policy formulation under conditions of greater climate variability are more challenging than ever. Learning from the experiences of diverse women, the poor, and indigenous groups is increasingly important in order to effectively collaborate with researchers, publics, and policymakers to find ways to ameliorate the negative effects of global environmental change. As plans and funding for adaptation emerge at international and local levels, some community-based adaptation projects address gender inequities and include local input, while other projects are large-scale infrastructural dams or coastal protections that employ mostly men or do not protect resources of importance to women (Tschakert and Machado 2012). Soma Chakrabarti (2014) reminds us that in formulating climate adaptation and resilience programs and policies, it is important to tackle women's workloads, which are often greater than men's and tend to increase as a result of climate change. Women's vulnerability can be reduced where they have strong social networks, are connected to assistance programs or alternative work opportunities, or use their traditional and embodied knowledge to cope with and create sustainable change (Tschakert 2012).

All of the chapters in this book have clear policy relevance; some also document policy change as a result of women's or indigenous people's influence in the policy process. The different studies presented offer approaches that help foster inclusivity in working toward more equitable, effective, and enduring policies. Hanson's contribution on women's recycling in coastal Yucatán shows that short talks by non-profit organizations or governmental organizations about waste management can blossom into community-led efforts such as the recycling cooperatives set up by women that spread throughout the coast. Hanson's case study strongly points also to the need for better waste management at the global level, especially waste reduction, but also on-site recycling. Reducing waste will need to be more widely promoted from the level of individuals and households to organizations, institutions, and businesses. As Margaret Mead stated (probably in a speech at the first Earth Day celebration in the U.S.): "Never doubt that a small group of thoughtful, concerned people can change the world. Indeed, it's the only thing that ever has" (quoted in Lutkehaus 2008: 261).

Perkins and Walker's and Moraes's contributions also focused on larger-scale impacts that women can have, but their cases examine the watershed scale. Perkins and Walker stressed that community engagement through sister watershed projects and trainings was particularly successful as a strategy to educate and to involve girls, boys, and women in understanding and acting on watershed protection in the context of a changing climate. Moraes's case study of federal-level programs in semi-arid northeast Brazil that promoted cistern building, including among women, illustrated that such programs can help reduce women's workload related to household water provision. These programs also improve the quantity and quality of the water the women can access in part by evening out the availability of water throughout the year even in the context of a changing precipitation regime. Her study points to the active participation of women in micro-levels of watershed management such as at the household and community scales but also to the barriers women—particularly poorer, less educated women and women of color—still face in participating in regional-level watershed management.

Hayman's chapter on Tlingit communities in the Yukon, Canada and Lahiri-Dutt's chapter on a village in the state of Jharkhand, India revealed that indigenous peoples have had creative methods to resist neo-colonial and capitalist domination of their land and water. Their narratives could be useful in projects and policies aimed at bringing these populations into the economy and into the political arena as active agents of their own destiny. Their interests include the protection of water sources in these two cases and as such can offer guidance to all groups seeking to protect natural resources in their surroundings. Williams and Golovnev's and Hayman's focus on indigenous populations helps to further policy, programmatic, and academic understandings of multi-scalar economic effects on indigenous women and men of climate and other ecological changes on water resources (including in the form of ice) necessary for their livelihoods that include farming, hunting, and fishing.

Several of the chapters illuminate lessons for program and policy formulation by documenting women's active participation in water-related challenges that effect the gathering of wild plants and herbs for food and medicinal purposes and the production and/or processing of food. The policy dimensions of Braun's work on highland Lesotho include the message that listening to local needs and privileging non-capitalist, local economies can benefit regions more in both the short and long term than tried and failed large-scale development projects like large dams. In the case of semi-arid Sonora, Mexico (Buechler) and the Nile river area of Egypt in the Mubarak Resettlement Scheme (Najjar), it was evident that understanding and incorporating women's adaptation strategies into agricultural policymaking can help reduce vulnerability and conserve water and fragile lands. Williams and Golovnev's study of the Pamir region revealed that women's creative strategies related to cropping and to livestock rearing are bearing fruit in terms of reducing vulnerability to climate change. These authors elucidate that sometimes the most effective tool to address environmental change impacts can be helping to share the voices and images of the social groups who are, on a daily basis, using their local knowledge and lived experience to make small adjustments to address climate and water changes.

Visions for the future of applied research and action on women, water, and global environmental change

In looking ahead to the future of applied research and action on women, water, and global environmental change, there is a clear need for multi-disciplinary theoretical lenses that continue to creatively bridge academic practice with the everyday activism and local, embodied actions of women within large-scale water and climate governance. Not just limited to women, the effects of intersecting social (race, ethnicity, age) and physical locations (such as the urban–rural continuum, community to watershed, and ice, lake, sea, and riparian areas or desert environments) also need to be incorporated into this work. Interdisciplinary and multi-scalar collaboration can more effectively demonstrate the persistent networks of power and marginalization linked to political, social, economic, and environmental issues. Political ecology and feminist frameworks help to underscore the intricacies of local priorities as well as macro-level challenges and opportunities in relation to global environmental change. As Gibson-Graham and Roelvink have argued:

> Small actions and networks can be seen to have sweeping global effects, and rapid large-scale change can emerge from diffuse local transformations. Theory has taken on a new relation to action—to understand the world is to change it. As a performative practice, academic research is activism; it participates in bringing new realities into being.
>
> (2009: 342)

In terms of this volume, we believe that the different case studies presented offer creative advances to theoretical, methodological, and practical approaches that help foster inclusivity in working toward more equitable, effective, and enduring policies. Paying attention to women's lives and actions, especially through a feminist lens, is essential to understanding environmental change and our responses to it, and to developing strategies that will sustain waterscapes and livelihoods around the world.

References

Agarwal, B. 2001. Participatory exclusions, community forestry, and gender: An analysis for South Asia and a conceptual framework. *World Development*. 29 (10): 1623–1648.

Ahlers, R. and M. Zwarteveen. 2009. The water question in feminism: Water control and gender inequities in a neo-liberal era. *Gender, Place & Culture*. 16 (4):409–426.

Aladuwaka, S. and J. Momsen. 2010. Sustainable development, water resources management and women's empowerment: The Wanaraniya Water Project in Sri Lanka. *Gender & Development*. 18 (1):43–58.

Alagan, R. and S. Aladuwaka. 2014. Gender mapping in post-disaster recovery. In *Global perspectives on gender and space: Engaging feminism and development*, eds A.M. Oberhauser and I. Johnston-Anumonwo, 181–191. London: Routledge.

Arnell, N.W. 2004. Climate change and global water resources: SRES emissions and socio-economic scenarios. *Global Environmental Change*. 14 (1):31–52.

Ávila Sanchez, H. 2005. *Lo urbano-rural, ¿nuevas expresiones territoriales?* Cuernavaca, Mexico: Universidad Nacional Autónoma de México, Centro Regional de Investigaciones Multidisciplinarias.

Bee, B. 2013. Who reaps what is sown? A feminist inquiry into climate change adaptation in two mexican ejidos. *ACME: An International E-Journal for Critical Geographies*. 12 (1):131–154.

——, M. Biermann, and P. Tschakert. 2013. Gender, development, and rights-based approaches: Lessons for climate change adaptation and adaptive social protection. In *Research, action and policy: Addressing the gendered impacts of climate change*, eds M. Alston and K. Whittenbury, 95–108. Dordrecht, Netherlands: Springer.

Blaikie, P. 2012. Should some political ecology be useful? The inaugural lecture for the cultural and political ecology specialty group, annual meeting of the Association of American Geographers, April 2010. *Geoforum*. 43 (2):231–239.

Braidotti, R., E. Charkiewicz, S. Hausler, and S. Wieringa. 1994. *Women, the environment and sustainable development*. London: Zed Books.

Braun, B. 2005. Environmental issues: Writing a more-than-human urban geography. *Progress in Human Geography*. 29 (5):635–650.

Brockington, D., R. Duffy, and J. Igoe. 2008. *Nature unbound: Conservation capitalism, and the future of protected areas*. London: Earthscan.

Buechler, S. 2002. Irrigated agriculture on Mexican ejidos: Complementarities with off-farm and non-farm economic strategies. In *Managing a sacred gift: Changing water management strategies in Mexico*, eds S. Whiteford and R. Melville, 125–142. La Jolla: Center for U.S.-Mexican studies, University of California-San Diego.

Chakrabarti, S. 2014. *The gender advantage: Women on the front line of climate change*. Rome: International Fund for Agricultural Development.

Cleaver, F. and K. Hamada. 2010. "Good" water governance and gender equity: A troubled relationship. *Gender & Development*. 18 (1):27–41.

Coles, A., L. Gray, and J. Momsen. 2015. *Routledge handbook of gender and development*. London: Routledge.

Combes, S. 2003. Protecting freshwater ecosystems in the face of global climate change. In *Buying time: A user's manual for building resistance and resilience to climate change in natural systems*, eds L. Hansen, J.L. Biringer, and J. Hoffman, 177–216. Berlin, Germany: World Wildlife Fund Climate Change Program. www.cakex.org/sites/default/files/buyingtime_unfe.pdf. Accessed August 15, 2014.

Cruz-Torres, M.L. and P. McElwee. 2012. *Gender and sustainability: Lessons from Asia and Latin America*. Tucson, AZ: University of Arizona Press.

Dankelman, I. 2002. Climate change: Learning from gender analysis and women's experiences of organising for sustainable development. *Gender & Development*. 10 (2):21–29.

Dyck, I. 2005. Suzanne Mackenzie memorial lecture. Feminist geography, the "everyday," and local–global relations: Hidden spaces of place-making. *The Canadian Geographer*. 49:233–243.

Elwood, S. and D. Martin. 2000. "Placing" interviews: Location and scales of power in qualitative research. *Professional Geographer*. 52:649–657.

Escobar, A. 2004. Beyond the Third World: Imperial globality, global coloniality and anti-globalisation social movements. *Third World Quarterly*. 25 (1):207–230.

Fortmann, L. 2008. *Participatory research in conservation and rural livelihoods: Doing science together*. London: Wiley-Blackwell.

GESEC. 2011. Feminist political ecology. In *The women, gender and development reader*, eds N. Visvanathan, L. Duggan, N. Wiegersma, and L. Nisonoff, 43–46. London: Zed Books.

Gibson-Graham, J.K. 1994. "Stuffed if I know": Reflections on post-modern feminist social research. *Gender, Place & Culture*. 1:205–224.

——. 2006. *A postcapitalist politics*. Minneapolis, MN: University of Minnesota Press.

—— and G. Roelvink. 2009. An economic ethics for the anthropocene. *Antipode*. 41 (1):320–346.

Giritli-Nygren, K. and U. Schmauch. 2012. Picturing inclusive places in segregated spaces: A participatory photo project conducted by migrant women in Sweden. *Gender, Place & Culture*. 19 (5):600–614.

Goldsmith, R., I. Fyegina, and J.T. Jost. 2013. The gender gap in environmental attitudes. In *Research, action and policy: Addressing the gendered impacts of climate change*, eds M. Alston and K. Whittenbury, 159–174. Dordrecht, Netherlands: Springer.

Harriss, R. 2014. Epilogue: The end of the Arctic as we know it. In ed. J. Norwine, *A world after climate change and culture-shift*, 393–403. Dordrecht, Netherlands: Springer.

Harvey, D. 1996. *Justice, nature and the geography of difference*: Malden, MA: Blackwell.

Herman-Mercer, N., P.F. Schuster, and K.T.B. Maracle. 2011. Indigenous observations of climate change in the Lower Yukon River Basin, Alaska. *Human Organization*, 70 (3):244–252.

Heynen, N., M. Kaika, and E. Swyngedouw. 2005. *In the nature of cities: Urban political ecology and the politics of urban metabolism*. Florence, KY: Routledge.

Hodson, M. and S. Marvin. 2010. Urbanism in the anthropocene: Ecological urbanism or premium ecological enclaves? *City*. 14 (3):298–313. doi: 10.1080/13604813.2010.482277.

Howarth, R.W., R. Santoro, and A. Ingraffea. 2011. Methane and the greenhouse-gas footprint of natural gas from shale formations. *Climatic Change*. 106 (4): 679–690.

Jiménez-Solares, C. 1998. Lo rural y lo urbano: una reflexion en torno a su definicion. *Sociologia Rural*.

Johnson, J.M., A.J. Franzluebbers, S.L. Weyers, and D.C. Reicosky. 2007. Agricultural opportunities to mitigate greenhouse gas emissions. *Environmental Pollution*. 150 (1):107–124.

Kaijser, A. and A. Kronsell. 2014. Climate change through the lens of intersectionality. *Environmental Politics*. 23 (3):417–433.

Katz, C. 2001. On the grounds of globalization: A topography for feminist political engagement. *Signs: Journal of Feminist Theory*. 26 (4):1213–1234.

Kearney, M. 1996. *Reconceptualizing the peasantry*. Boulder, CO: Westview Press.

Keil, R. 2003. Urban political ecology. *Urban Geography*. 24 (8):723–738.

Khamis, M., T. Plush, and C.S. Zelaya. 2009. Women's rights in climate change: Using video as a tool for empowerment in Nepal. *Gender and Development*. 17(1): 125–135.

King, E. 2012. Are women the key to moving climate change debate forward? Podcast. Rttc.org. To hear podcast go to: www.rtcc.org/2012/06/29/are-women-the-key-to-moving-climate-change-debate-forward/#sthash.6K9KwG4s.dpuf. Accessed July 29, 2014.

Lankao, P. and J.B. Smith (coordinating authors). 2014. Chapter 26. North America. Intergovernmental Panel on Climate Change (IPCC). Fifth Assessment Report. www.ipcc.ch/report/ar5/. Accessed June 15, 2014.

López Ruelas, L. 2014. Acid spill in river disrupting Sonorans. *Arizona Daily Star*. A2; A7. August 20.

Lutkehaus, N. 2008. *Margaret Mead: The making of an American icon*. Princeton, NJ: Princeton University Press.

McCright, A.M. 2010. The effects of gender on climate change knowledge and concern in the American public. *Population and Environment*. 32:66–87.

McDonald, R.I., R.T.T. Forman, P. Kareiva, R. Neugarten, D. Salzer, and J. Fisher. 2009. Urban effects, distance, and protected areas in an urbanizing world. *Landscape and Urban Planning*. 93 (1):63–75.

Milman, O. 2014. Entire marine food chain at risk from rising CO2 levels in water. *The Guardian*, April 13.

Mollett, S. and C. Faria. 2013. Messing with gender in feminist political ecology. *Geoforum*. 45:116–125. doi: 10.1016/j.geoforum.2012.10.009.

Momsen, J. 2010. *Gender and development*. London: Routledge.

Nagar, R. 2006. *Playing with fire: Feminist thought and activism through seven lives in India*. Minneapolis, MN: University of Minnesota Press.

—— and F. Ali. 2003. Collaboration across borders: Moving beyond positionality. *Singapore Journal of Tropical Geography*. 24:356–372.

Nightingale, A. 2003. A feminist in the forest: Situated knowledges and mixing methods in natural resource management. *ACME: An International E-Journal for Critical Geographies*. 2 (1):77–90.

——. 2011. Bounding difference: Intersectionality and the material production of gender, caste, class and environment in Nepal. *Geoforum*. 42 (2):153–162. doi: 10.1016/j.geoforum.2010.03.004.

Nijhuis, M. 2014. Can coal ever be clean? *National Geographic Magazine* (April): 28–61.

Osnes, B. 2014. *Theatre for women's participation in sustainable development*. London: Routledge.

Petherick, A. 2013. Oblivious money-men. *Nature Climate Change*. 3:96–97.

Pillow, W.S. and C. Mayo. 2011. Feminist ethnography. In *Handbook of feminist research: Theory and praxis*, ed. S.N. Hesse-Biber, 187–205. Thousand Oaks, CA: Sage.

Rocheleau, D. 1995. Maps, numbers, text, and context: Mixing methods in feminist political ecology. *The Professional Geographer*. 47 (4):458–466. doi: 10.1111/j.0033-0124.1995.458_h.x.

Salud Sonora. 2014. Estimación de población total 2014 por jurisdicción sanitaria, coordinación médica local y municipio. Gobierno de México.

Sankey, K. and I. Cárdenas. 2013. Mujeres y crisis de reproducción social: La otra crisis del agua. *Observatorio del Desarrollo*. 2 (7):10–15.

Seager, J. 2009. Death by degrees: Taking a feminist hard look at the 2 degrees climate policy *Kvinder, Køn & Forskning*. 18 (34):11–22.

Shanley, P. 2006. Science for the poor: How one woman challenged researchers, ranchers, and loggers in Amazonia. *Ecology and Society*. 11 (2):28.

Sharp, J. 2005. Geography and gender: Feminist methodologies in collaboration and in the field. *Progress in Human Geography*. 29:304–309.

Stewart, A.J. and E.R. Cole. 2007. Narratives and numbers: Feminist multiple methods research. In *Handbook of feminist research: Theory and praxis*, ed. S.N. Hesse-Biber, 327–344. Thousand Oaks, CA: Sage.

Sundberg, J. 2004. Identities-in-the-making: Conservation, gender, and race in the Maya biosphere reserve, Guatemala. *Gender, Place and Culture*. 11 (1):43–66.

Townsend, J.G. 2005. *Women's voices from the rainforest*. Second ed. London: Routledge.

Truelove, Y. 2011. (Re-)conceptualizing water inequality in Delhi, India through a feminist political ecology framework. *Geoforum*. 42 (2):143–152. doi: 10.1016/j.geoforum.2011.01.004.

Tschakert, P. 2012. From impacts to embodied experiences: Tracing political ecology in climate change research. *Geografisk Tidsskrift-Danish Journal of Geography*. 112: 144–158.

—— and M. Machado. 2012. Gender justice and rights in climate change adaptation: Opportunities and pitfalls. *Ethics and Social Welfare*. 6:275–289.

Tuana, N. and C.J. Cuomo. 2014. Climate change—Editors' introduction. *Hypatia*. doi: 10.1111/hypa.12088.

UNFCCC. 2014. Gender and climate change. http://unfccc.int/gender_and_climate_change/items/7516.php. Accessed August 9, 2014.

Visvanathan, N., L. Duggan, N. Wiegersma, and L. Nisonoff. 2011. *The women, gender and development reader*. Second ed. New York: Zed Books.

Walker, D., J.P. Jones, III, S.M. Roberts, and O.R. Frohling. 2007. When participation meets empowerment: The WWF and the politics of invitation in the Chimalapas, Mexico. *Annals of the Association of American Geographers*. 97 (2):423–444.

Walker, P.A. 2006. Political ecology: Where is the policy? *Progress in Human Geography*. 30 (3):382–395.

Wamukonya, N. and M. Skutsch. 2002. Gender angle to the climate change negotiations. *Energy and Environment*. 13:115–124.

Webb, M.D. and K.W. Howard. 2011. Modeling the transient response of saline intrusion to rising sea-levels. *Groundwater*. 49 (4):560–569.

Weichselgartner, J. and R. Kasperson. 2010. Barriers in the science-policy-practice interface: Toward a knowledge-action-system in global environmental change research. *Global Environmental Change*. 20 (2):266–292.

Williams, R.W. 1973. *The country and the city*. Oxford and New York: Oxford University Press.

Zaman, F. and J. Reichert. 2014. *This time next year*. A film on rebuilding after Hurricane Sandy produced for the resilience initiative of the Rockefeller foundation. www.rockefellerfoundation.org/blog/this-time-next-year. Accessed August 12, 2014.

Appendix

Tlingit/Tagish stories

From: Eleanor Hayman Chapter 10

Community consultant Colleen James was quite clear that this story about Crow/*Yéil* should be one of the key stories about water included in this book because it tells the story of how the many lakes in the region were formed and how the fish appeared in the lakes. Many other CTFN community members also told me various versions of this 'Crow steals water' story as did David Katzeek. This version was told by Angela Sidney in 1950 and recorded by Catharine McClellan.

'How Crow Got Water' (Version One)

September 27, 1950, Carcross

Crow went outside.
It was the wintertime, and he got some frozen excrement and brought it back inside.
He dropped some under the man [who had the water].
Soon it began to thaw out and smell.
Crow says, "ax súnee (my uncle, father's clan), what's that smell? Look what you've done. You've done something to yourself!"
The old man says, "How come I did that? Maybe I stepped on it."
So he went outside to clean himself up.
And Crow drank as much water as he could. He almost burst.
When the old man started coming in, Crow flew up through the smokehole.
And Crow always makes a noise when he flies up.
The old man was mad.
He said to his powers of the roof hole, "Grab him!"
So Crow got stuck. He tries to fly, and he can't fly.
The old man gets dry pitch and smokes him. [Angela explained that this dry pitch makes good kindling.]
Finally Crow is almost choked, but he gets away.
Then he drops down a little ways. He is so full, he can't move.
He takes a good rest, and then he starts to fly, and he drips fishes and lakes all over the place.

In: *My Old People's Stories: A Legacy for Yukon First Nations*. 2007. 3 vols. Julie Cruikshank, ed. and additions. Catherine Kernan, illustrations. Occasional Papers in Yukon History 5 (1–3): 257. Whitehorse: Government of Yukon, Cultural Services Branch.

'The Man Behind the Dam' (in Tlingit: *Héidu dei k̲áa*) is a Deisheetaan (Beaver) Clan story rooted in the time when the coastal Tlingit were first exploring the Yukon interior. As community consultant Mark Wedge stresses, it is more importantly a story about water and the sacred. This is a much shortened version of Mark Wedge's telling from an interview in August 2013.[1]

'The Man Behind the Dam'

A coastal Tlingit man named Naatz comes inland to trap beaver. This is done by setting a beaver net hung with moose hoof bells under the ice on lakes and rivers. However, Naatz is unlucky. Naatz notices a wolverine following him and asks what it wants. The wolverine replies by telling Naatz to set his net where the water flows over the dam – not the usual procedure for trapping beaver. The wolverine further cautions 'You're going to catch something but what you catch is not for you to keep. You have to learn what it has to teach you then set it back in the water (*héen*).' Naatz does as he is told and immediately the moose hoof bells on the net start ringing. Naatz pulls up from the net in the water (*héen*) a beaver's head[2] with gold eyes and gold eyebrows, six legs and two tails. This is *Héidu dei k̲áa – the man behind the dam*.

Naatz continues to set beaver nets successfully. Naatz loads up his caribou shank toboggan with all the beaver hides and sets off back to the coast. On his way he sees porcupine tracks and as he is hungry, follows them into a cave. As Naatz enters the cave a huge stone rolls to cover the exit. Frantic, Naatz tries to move the stone and find another way out. He fails and becomes increasingly exhausted and depressed, so much so that he lies down on the cave floor. A small mouse runs to him and whispers in Naatz's ear '*You use your head. You've got everything it takes to get you out of here.*' Naatz fumbles around and finding porcupine droppings gathers and lights them with his flint. The resulting fumes create such a stink that voices chastise him and the stone rolls back from the cave mouth.

Naatz escapes back to his caribou shank toboggan to find it and the wolverine deeply covered in fresh snow. Naatz has no idea how long he has been in the cave but now continues his way home to the coast.

Notes

1 A version of this story is recorded by Catharine McClellan in *My Old People Say: An Ethnographic Survey of the Southern Yukon Territory*, Part II, pp. 471–473, 1975.
2 Mark says in another version it is a human head.

Index

Page references in bold type refer to figures and tables.